Design of Network Coding Schemes in Wireless Networks

Design of Network Coding Schemes in Wireless Networks

Zihuai Lin

CRC Press
Taylor & Francis Group
Boca Raton London New York

CRC Press is an imprint of the
Taylor & Francis Group, an **informa** business

First edition published 2022
by CRC Press
6000 Broken Sound Parkway NW, Suite 300, Boca Raton, FL 33487-2742

and by CRC Press
2 Park Square, Milton Park, Abingdon, Oxon, OX14 4RN

© 2022 Zihuai Lin

CRC Press is an imprint of Taylor & Francis Group, LLC

Reasonable efforts have been made to publish reliable data and information, but the author and publisher cannot assume responsibility for the validity of all materials or the consequences of their use. The authors and publishers have attempted to trace the copyright holders of all material reproduced in this publication and apologize to copyright holders if permission to publish in this form has not been obtained. If any copyright material has not been acknowledged please write and let us know so we may rectify in any future reprint.

Except as permitted under U.S. Copyright Law, no part of this book may be reprinted, reproduced, transmitted, or utilized in any form by any electronic, mechanical, or other means, now known or hereafter invented, including photocopying, microfilming, and recording, or in any information storage or retrieval system, without written permission from the publishers.

For permission to photocopy or use material electronically from this work, access www.copyright.com or contact the Copyright Clearance Center, Inc. (CCC), 222 Rosewood Drive, Danvers, MA 01923, 978-750-8400. For works that are not available on CCC please contact mpkbookspermissions@tandf.co.uk

Trademark notice: Product or corporate names may be trademarks or registered trademarks and are used only for identification and explanation without intent to infringe.

ISBN: 9781032067766 (hbk)
ISBN: 9781032067780 (pbk)
ISBN: 9781003203803 (ebk)

DOI: 10.1201/9781003203803

Typeset in NimbusSanL-Regu font
by KnowledgeWorks Global Ltd.

Publisher's note: This book has been prepared from camera-ready copy provided by the authors.

Contents

List of Figures ix

List of Tables xiii

Acknowledgments xv

Acronyms xvii

Author Biography xxi

Chapter 1   Introduction 1
                1.1   Physical Layer Lattice Network Coding and Soft Information Delivery 2
                      1.1.1   Lattice coding 2
                      1.1.2   PNC soft information and delivery 2
                1.2   Network Layer Network coding schemes 4
                      1.2.1   XOR network coding 4
                      1.2.2   Linear network coding 4
                              1.2.2.1   Random linear network coding 5
                              1.2.2.2   Distributed random linear network coding 6
                      1.2.3   Benefits made by network coding 6
                              1.2.3.1   Bandwidth efficiency 6
                              1.2.3.2   The basic idea 7
                              1.2.3.3   Other traffic configurations 8
                              1.2.3.4   Undirected networks 8
                      1.2.4   Energy efficiency 8
                              1.2.4.1   Multicast 9
                              1.2.4.2   Other traffic configurations 9
                      1.2.5   Delay performance 9
                              1.2.5.1   The average delay 10
                              1.2.5.2   Delay distribution 10
                      1.2.6   Reliability 11
                              1.2.6.1   Retransmissions and network coding 11
                              1.2.6.2   Combination of routing and network coding 13

	1.3	Network coding design challenges 14
	1.4	Organization of the Book 14
Chapter 2		Wireless Network Coded Systems for Multiple Interpretations 25
	2.1	Introduction 25
	2.2	System Model 26
	2.3	Optimization formulation 28
	2.4	Analysis of the Average Channel Capacity 29
	2.5	Network Coded System based on Nested Codes 32
		2.5.1 Soft-Decision Decoding with Nested Codes 34
	2.6	Analytical Bounds on the Bit Error Probability 35
	2.7	Code Search 37
	2.8	Numerical and Simulation Results 38
		2.8.1 Average Channel Capacity and Outage Probability 38
		2.8.2 The Performance of OS 42
		2.8.3 The Performance of Nested Codes 43
	2.9	Conclusions 44
Chapter 3		Distributed Network Coded Modulation Schemes for Multiple Access Relay Channels 47
	3.1	Introduction 47
	3.2	System Model 48
	3.3	Distributed Network Coded Modulation Schemes based on Punctured Convolutional Codes 48
		3.3.1 Decoding with Network Coded Modulation at the Destination Node 51
		3.3.2 Analytical bounds on the bit error probability for the multiple access relay channels 53
	3.4	Interleaved Distributed Network Coded Systems 57
	3.5	Simulation Results for Distributed Network Coded Systems 59
		3.5.1 Simulation results for Distributed Network Coded System without Interleaver 59
		3.5.2 Simulation results for Interleaved Distributed Network Coded System 60
	3.6	Summary 61
Chapter 4		Lattice Network Coding for Multi-Way Relaying Systems 67
	4.1	Introduction 67
	4.2	System Model 68
		4.2.1 System Model 68
		4.2.2 Nested Convolutional Codes and Lattice Network Coding 68
	4.3	Nested Convolutional Lattice Network Codes 70

	4.4	Performance Analysis 74
	4.5	Numerical Simulation Results 77
	4.6	Conclusion 78

Chapter 5 Nested LDGM-based Lattice Network Codes for Multi-Access Relaying Systems 81

    5.1 Introduction 81
    5.2 System Model 82
    5.3 Coding Process: Nested Binary LDGM Codes 83
    5.4 Coding Process: Nested Non-binary LDGM with Lattice 86
    5.5 L-EMS Decoding Algorithm 90
    5.6 Performance Analysis 91
    5.7 Code Optimization using Lattice based Monte Carlo Method 93
    5.8 Numerical and Simulation Results 97
        5.8.1 Lattice Settings 97
        5.8.2 Lattice-based Monte Carlo Method 97
        5.8.3 Performance for the Lattice-based EMS decoder 98
        5.8.4 Performance of the nested non-binary LDGM codes with lattice 99
    5.9 Conclusion 100

Chapter 6 Design of Soft Network Coding for Two-Way Relay Channels .. 103

    6.1 Introduction 103
    6.2 System Model 104
    6.3 TCQ codebook Design 106
    6.4 Performance Analysis 108
        6.4.1 Set a threshold 108
        6.4.2 Performance Analysis on the Two schemes 109
    6.5 Simulation Results 110
    6.6 Conclusion 112

Chapter 7 Linear Neighbor Network Coding 113

    7.1 Introduction 113
    7.2 System Model 114
    7.3 Theoretical Analysis 114
        7.3.1 Construction of the States 115
        7.3.2 Transition Matrices 115
        7.3.3 The State Vectors 117
        7.3.4 Reliability 118
        7.3.5 Networks without Network Coding 118
    7.4 Bounds on the Reliability 119
        7.4.1 The Upper Bound 119
        7.4.2 The Lower Bound 120

	7.5	Results and Discussion 121
	7.6	Conclusions 124
Chapter 8		Random Neighbor Network Coding 127
	8.1	Introduction 127
	8.2	System model 127
	8.3	Theoretical analysis 128
		8.3.1 States 129
		8.3.2 Transition matrices 130
		8.3.3 Probability vector and the reliability 134
	8.4	Optimisations 134
		8.4.1 Optimize the reliability at an individual round 135
		8.4.2 Optimize the expected round to absorb 135
	8.5	Numerical results 135
		8.5.1 Validation of the theoretical analysis 136
		8.5.2 Optimal selection of the tuning parameter 136
		8.5.3 Examination on the reliability gain 139
		8.5.4 Comparison with the random linear network coding scheme 140
	8.6	Summary 141
Index		 143

List of Figures

1.1 XOR network coding 4
1.2 Illustration of the column vector assignment to different channels. 5
1.3 Traditional butterfly networks with routing or network coding schemes, where N_1 sends x_1 to N_4 and N_2 transmits x_2 to N_6 7
1.4 A network with five nodes using a combined network coding and retransmission scheme. 12
1.5 Routing vs Network Coding. 13

2.1 A network coding group with four source nodes and four destination nodes and one relay node. 27
2.2 The coding process of a nest network coded system 32
2.3 Opportunistic scheduling against fixed scheduling with $N_T = 5$, $k = 1, 2$ and rate adaptation with/without optimal power allocation. 40
2.4 Opportunistic scheduling against fixed scheduling with $N_T = 10$, $k = 1, 2$ and rate adaptation with/without optimal power allocation. 41
2.5 Outage probability for opportunistic and fixed scheduling for N = 5,10, k = 1,2 with optimal power and rate adaptation. 41
2.6 The bit error probability performance of the received XORed packets before the last decoder at destination nodes under different OS selections. 42
2.7 The average number of the received packets per slot at different destination nodes under OS. 43
2.8 The performance of nested codes with different side information when OS level $k = 1$, RCPC code rate is 1/3. 44

3.1 An uplink transmission system with one source MT, a cooperative MT, a Relay node and a BS. 49
3.2 Iterative decoding for interleaved distributed network coded systems. 58
3.3 Analytical upper bounds and one simulation result for punctured trellis coded MSK (i.e., CPFSK with 1REC and $h = 1/2$) with $r = 1/2, 2/3, 3/4$, $G(D) = [1, D + D^2]$ and $G(D) = [1 + D^2; 1 + D + D^2]$. 60
3.4 EXIT chart for interleaved distributed network coded MSK (i.e., CPFSK with 1REC and $h = 1/2$) with $r = 2/5, 1/2, 2/3, 3/4$, and $G(D) = [1 + D + D^2; 1 + D^2]$. 62
3.5 BER performance for interleaved distributed network coded MSK (i.e., CPFSK with 1REC and $h = 1/2$) with $r = 2/5, 1/2, 2/3, 3/4$, $G(D) = [1 + D + D^2; 1 + D^2]$. 63

4.1 Multiple source nodes and single relay. 68
4.2 NCLC procedure at the relay node and the source node s_j 70

4.3	Upper bound on the WER and the simulation results for the investigated system.	77
5.1	The compute-and-forward model.	82
5.2	A special case with three sources, one relay, and one destination.	84
5.3	Probability distribution function of the received signal with three source nodes.	84
5.4	Code performance at relay and destination nodes with degree 6.	85
5.5	Procedures of the non-binary nested LDGM with lattice at the relay and destination nodes.	87
5.6	A new coordinate system for a practical Gaussian integer lattice constellations	95
5.7	A practical Gaussian integer lattice constellations with the message space $\mathbb{F}_{13} \cong \mathcal{W}$, and $\mathcal{W} \cong \mathbb{Z}[i]/\delta\mathbb{Z}[i]$, where $\delta = 2 + 3i$	97
5.8	$\overline{w}_c$ optimization in the case of code length $N = 2 \times 10^3$, rate $R = 0.5$, and finite field $\mathbb{F}_{13}$.	98
5.9	Comparison of Lattice-based EMS decoding performances for the different n_m values over finite filed $\mathbb{F}_{13}$.	99
5.10	The Monte-Carlo simulation result of the WER of the message at the destination in the compute-and-forward scheme with three transmitters, one relay and one destination.	100
6.1	A two-way relay network.	105
6.2	A two-way relay network.	106
6.3	A $(5,2)_8$ convolutional encoder and its trellis diagram.	107
6.4	QPSK constellation.	108
6.5	BER performance for fading channels.	111
6.6	BER performance for fading channels.	111
7.1	Illustration of a lossy wireless network with n nodes applying neighbor network coding. Note that some paths are not shown in the figure.	115
7.2	Simulation and theoretical results of the reliability of networks when $n = 3, 4, 5$, where the probabilistic connectivity matrix is given in eq. 7.18.	122
7.3	The reliability gain of neighbor coding network over non-coded network when $n = 4$.	123
7.4	The reception of x_1 by the network with different neighbor network coding schemes, where the connectivity matrices are given in Eq. 7.19.	123
7.5	Bounds on the probability that N_3 receives x_1 where $n = 3$ nodes and the connectivity matrix is given by 7.20.	124
8.1	A state transition diagram for a wireless network with three nodes.	131
8.2	Theoretical and simulation results comparison for the network reliability applying RNNC when n = 3 and 4. The probabilistic connectivity matrices are given by Eq. (8.19).	136

List of Figures

8.3 The reliability for a three node network at the 4th round when ω varies form zero to one, the probabilistic connectivity matrix is given by eq. (8.20)..137

8.4 The reliability for a three node network at the 6th round when ω varies form zero to one, the probabilistic connectivity matrix is given by eq. (8.20)..138

8.5 The reliability comparison for networks with three and four nodes. The connectivity matrices are given in eq. (8.23)..139

8.6 Performance comparison for different network coding schemes. The probabilistic connectivity matrix is given by Eq. (8.24). 140

List of Tables

2.1 Table of good codes 37
2.2 Distance spectrum for code rates $1/6$ and $2/6$ 38
2.3 Distance spectrum for code rates $3/6$ and $4/6$ 39

3.1 Punct. matrices for the investigated Punctured trellis coded CPFSK schemes, $h = 1/2$, 1REC. 59
3.2 Puncturing and repetition matrices for the investigated interleaved distributed network coded systems. 61

7.1 The states of N_1 and corresponding packets for a network of three nodes, with N_3, N_1, and N_2 being the coding neighbours for N_1, N_2, and N_3, respectively. The 6th state, for example, is $\{111\}$, which indicates that N_1 has packets x_1, x_2 and x_3. 116

8.1 Optimal ω values for a three node network at different round with the probabilistic connectivity matrix given by Eq. 8.20. 138

Acknowledgments

A list of people who assisted me and made it possible to finish the work would be very long but space forbids me to mention them all by name. However, I would like to thank all the former and present colleagues and friends at the School of Electrical and Information Engineering, University of Sydney for the contribution of the nice and friendly atmosphere here. Specially, I would like to thank Dr. Jing Yue, Dr. Yunaye Ma, Dr. Jun Li, Ms. Li Ma, Ms. Yiwen Li, Dr. Peng Wang, Dr. Ming Ding, Dr. Youjia Chen, Dr. Chuan Ma, Dr. Di Zhai, Dr. Xiaopeng Wang, Prof. Guoqiang Mao, Prof. Yonghui Li and Prof. Branka Vucetic for their enthusiastic support and invaluable discussions. Their patience and time are very much appreciated. My thanks also go to my present colleagues and friends at the Centre of IoT and Telecommunications and all other friends whose name are not mentioned here, thanks all of you for your friendship.

Zihuai Lin
Sydney, July 2021

Acronyms

AF	Amplify-and-Forward
AMC	adaptive modulation and coding
AP	Access Point
APP	*a posteriori* Probability
AR	AutoRegressive
ARQ	Automatic Repeat reQuest
AWGN	Additive White Gaussian Noise
BCJR	Bahl, Cocke, Jelinek and Raviv
BPSK	Binary Phase Shift Keying
BS	Base Station
BSC	Binary Symmetric Channel
CDF	Cumulative Distribution Function
CPE	Continuous Phase Encoder
CPFSK	Continuous Phase Frequency Shift Keying
CPM	Continuous Phase Modulation
CRC	Cyclic Redundancy Check
CSI	Channel State Information
CSNR	Channel Signal to Noise Ratio
DF	Decode and Forward
DPSK	Differential Phase Shift Keying
DNCC	Distributed Network-Channel Codes
EF	Estimate-and-Forward
EMS	Extended Min-Sum
EXIT	EXtrinsic Information Transfer
FEC	Forward Error Correction
FP	Fixed Power
FS	Fixed Scheduling
i.i.d.	independent, identically distributed
JDCE	Joint Distributed punctured Convolutional Encoder
JSCC	Joint Source and Channel Coding
JSCD	Joint Source and Channel Decoding
L-EMS	Lattice based Extended Min-Sum
LLR	Log-Likelihood Ratio
LDGM	Low Desity Generator Matrix
LDPC	Low-Density Parity-Check
LNC	Lattice Network Codes

MAC	Medium Access Control	
MAP	Maximum *a posteriori*	
MI	Mutual Information	
MIF	Mutual Information based Forwarding	
ML	Maximum Likelihood	
MLSD	Maximum Likelihood Sequence Detection	
MM	Memoryless Modulator	
MMSE	Minimum Mean Square Error	
MPEG	Moving Picture Experts Group	
MSE	Mean Square Error	
MSK	Minimum Shift Keying	
MT	Mobile Terminal	
MWRC	Multi-Way Relay Channel	
NC	Network Coding	
NCLC	Nested Convolutional Lattice Code	
NE	Network Encoder	
NLP	Nearest Lattice Point	
NSED	Normalized Squared Euclidean Distance	
ODS	optimum distance spectrum	
OS	opportunistic scheduling	
PAM	Pulse Amplitude Modulation	
PCC	Punctured Convolutional Code	
PDF	Probability Density Function	
PID	principle ideal domain	
PNC	Physical-layer Network Coding	
PSD	Power Spectral Density	
PSK	Phase Shift Keying	
QAM	Quadrature Amplitude Modulation	
QPSK	Quaternary Phase Shift Keying	
RCC	Rate Compatible Convolutional	
RCPC	Rate Compatible Punctured Convolution	
REC	RECtangular	
RN	Relay Node	
RNNC	Random Neighbor Network Coding	
RSC	Recursive Systematic Convolutional	
SCCPM	Serially Concatenated CPM	
SENR_norm	normalized signal-to-effective-noise ratio	
SER	Symbol Error Rate	
SIF	Soft information forwarding	

SISO	Soft In Soft Out
SNR	Signal to Noise Ratio
TCCPM	Trellis Coded Continuous Phase Modulation
TCQ	Trellis Coded Quantization
TCM	Trellis Coded Modulation
TWRC	Two-Way Relay Channels
WER	Codeword Error Rate

Author Biography

Zihuai Lin received the Ph.D. degree in Electrical Engineering from Chalmers University of Technology, Sweden, in 2006. Prior to this he has held positions at Ericsson Research, Stockholm, Sweden. Following Ph.D. graduation, he worked as a Research Associate Professor at Aalborg University, Denmark. At the same time, he worked at the Nokia Siemens Networks research center as an external senior researcher on 4G LTE standardization. He is currently a senior lecturer at the School of Electrical and Information Engineering, the University of Sydney, Australia. He has published more than 200 papers in international conferences and journals, which have been cited more than 2500 times. He holds twelve CN, three US, and one AU patents on LTE system design, distributed network coding and wireless sensor networks, microwave Ghost imaging, indoor localization, and ECG/EEG AI data analysis. His research interests include source/channel/network coding, coded modulation, massive MIMO, mmWave/THz communications, radio resource management, cooperative communications, small-cell networks, 5G/6G, IoT, wireless Artificial Intelligence (AI), ECG and EEG AI signal analysis, Radar imaging, etc.

1 Introduction

Network coding is a coding approach which can be utilized to enhance network capacity in comparison with conventional route procedure in a computer communication network [1–3]. Traditionally, routing is used in a classical communication network, e.g., internet, for information packets delivery, in which the received packets by the intermediate (relay) nodes between the source nodes and the destination nodes are simply stored and forwarded. Network coding protocols, however, allow the relay nodes to combine the received packets from different nodes for information delivery. It is shown in [1] that the maximum network throughput provided by the max-flow min-cut flow bound can be achieved by using network coding for the multicast case.

Network coding is proposed in [1] to solve the problem of defining admissible coding rate region, in which by using blocks code, the decoding error probability can be arbitrarily small. It reveals that, in general, multicasting information by store-and-forward is not an optimal solution. Rather, by combining and coding the received information at a relay node using network coding, bandwidth efficiency can be improved. In [1], directed graphs with error-free edges are used to represent the networks. Wired networks where links are usually lossless, e.g., an Internet backbone, can be described by this model. In [4, 5], linear network coding schemes are investigated which show that the network coding schemes can improve the network capacity. These works form the basis of network coding theory.

On the other hand, the network coding techniques applied in other network scenarios, such as, networks with transmission errors and undirected networks, are also being investigated. In a lossy network, information transmission errors cannot be avoided due to factors such as channel fading, interference, or mobility of devices. In this case, reliable transmission techniques, such as channel coding and network coding, are needed to protect the propagation errors. In [2, 3, 6], network error correction coding techniques are developed to solve the problem. For an error correction network code, if all error patterns less than or equal to the maximum number t of errors can be corrected, then we call that code having t error correction capability. Further works on error correction network codes can be found in [7, 8] for performance bounds, coding and decoding methods [9–11].

It's worth noting that the aforementioned study focuses on network coding in directed networks. In [12], the network coding approach is applied to undirected networks for the first time, demonstrating the capability to improve network throughput. The bounds on throughput are further investigated in [14–16] as a result of this work.

Although network coding was originally proposed for use at the network layer, its numerous advantages have prompted academics to consider using it at other protocol layers. The electromagnetic (EM) waves transmitted by different source nodes can be combined directly at a relay node over the air. As a result, rather than treating one

EM wave while regarding others as interference, multiple EM waves can be mixed to generate an output signal [17]. Such a coding scheme is known as the physical layer netowrk coding (PNC) scheme. Performance in terms of throughput, dependability, and latency of the PNC system have been studied in [17–19]. The advantages of PNC are proven by evaluating and comparing system performance in terms of the bit error rate and throughput of various transmission systems in [20] for a two-way relay network.

Other issues with network coding in terms of the size, the number of nodes for encoding and computing complexity have been studied in e.g., [13, 21, 22], etc. In addition to developing a theory of network coding, the application is expanded from multicast to other performance and other traffic configurations [23, 24].

1.1 PHYSICAL LAYER LATTICE NETWORK CODING AND SOFT INFORMATION DELIVERY

1.1.1 LATTICE CODING

Lattice codes have been used comprehensively to enhance spectrum efficiency in network coding schemes [25–28]. The lattice code pattern enables simultaneous transmissions from multiple sources to a relay with multi-user interference, resulting in high spectrum efficiency. [25] proposes a compute-and-forward strategy for achieving significantly higher rates by leveraging user interference.

Instead of treating the interference as noise, the relays use lattice codes to decode the linear combinations of transmitted signals into integer combinations of codewords. Based on the PNC [28] schemes, [27] develops a general algebraic framework called lattice network coding. The compute-and-forward strategy is reinterpreted in a generalized construction in the lattice network coding scheme, resulting in a type of linear network coding over modules. In this book, nested codes are reconstructed using lattices to achieve high spectrum efficiency. In Chapter 4, we investigate a multi-way relay channel (MWRC) [29–31], in which all users exchange information via a relay node. For the MWRC model, we develop nested convolutional lattice codes (NCLC) over a finite field enabling multiple interpretations for each user in two time slots.

Although the proposed NCLC improves spectrum efficiency, the decoding complexity grows exponentially as lattice dimensions increase. To reduce the coding and decoding complexity, non-binary low-density generator matrix (LDGM) codes are used in Chapter 5 to construct the lattice network codes. It has been shown that the complexity can be significantly reduced.

1.1.2 PNC SOFT INFORMATION AND DELIVERY

To improve the capacity of multiple access relay channels, network coding has been employed in cooperative relay networks [33, 34], and protocols aimed at these networks are known as PNC protocols [35, 36]. In PNC protocols, intermediate nodes directly combine analog signals from various source nodes [20, 36].

In PNC, the amplify-and-forward (AF) [37] and decode-and-forward (DF) [38] relay protocols have been extensively studied. The AF protocol amplifies the incoming signal and forwards it to the sink node, causing noise amplification at the relay; the DF protocol, on the other hand, first decodes the received signal, and then forwards the re-encoded signal to the sink node. It has the disadvantage of propagating erroneous decisions to the destination.

As a result, a new relaying technique known as soft information forwarding (SIF) was proposed to improve error performance by passing intermediate soft decisions at the relay [39–42]. Log-likelihood ratio (LLR) [39], soft symbol [40], or soft mutual information [41, 42] are examples of soft information.

These SIF protocols are, however, examined in single-way or two-way relay channels that do not apply directly to multi-source general relay channels, when multiple source nodes communicate with the sink node by relay. In wireless networks, such as wireless sensor networks, multiple access relay channels (MARC) are frequent and basic building pieces. Wireless network coding has been used in the MARC to improve network performance [43, 44]. Furthermore, it is usual in WSN for two source nodes in the MARC to have correlated information [45, 46].

With NC at the relay, redundant information will be contained in the network coded symbols of the correlated sources which can be compressed before transmitting the information to the destination. However, only unrelated sources are taken into account by present SIF protocols. Then it becomes a worthy research problem to effectively reduce redundancies in network coded signals at relay nodes.

The relay node produces the soft symbols in the soft-symbol based SIF protocol [40], also known as estimate-and-forward (EF), by calculating the minimal mean squared errors (MMSE) estimations of the received symbols at the relay. The EF protocol is shown to be able to maximize the signal-to-noise (SNR) at the destination, thus achieving better error performance than the AF and DF protocols.

Mutual information forwarding (MIF) [41, 42, 47, 48] is another recent soft information relaying protocol in which relay nodes forward symbol-wise mutual information (SMI) to the sink node. Results in [47] show that the MIF protocol outperforms the EF protocol in terms of error performance.

Relay protocol research has expanded to include increasingly complex wireless networks, such as those with two-way relay channels. This model has recently received a lot of attention because of its potential application in future wireless cooperation systems [41, 49–52].

Over fading two-way relay channels, all of these SIF protocols have been shown to outperform both AF and DF procedures. However, the SIF protocols make the assumption that the sink node perfectly know the soft information at relay nodes. This isn't practicable for real-world systems since the channel between the relay and sink nodes is band-limited and continuous real numbers necessitate more data, which isn't feasible in a network with limited channel capacity. Apart from that, even if such soft information can be quantized into a few bits, the SIF protocols appear to be defective when compared to the standard DF and AF protocols.

1.2 NETWORK LAYER NETWORK CODING SCHEMES

Network layer network coding schemes deal with data transmissions in the packet level. In the following, we will review several coding schemes in network layer.

1.2.1 XOR NETWORK CODING

Bitwise XOR is conducted among all or a specific set of packets at a node in XOR-based network coding systems. Consider the scenario depicted in Fig. 1.1 in which node N_1 wants to transmit its own packet x_1 to node N_2, and N_2 transmits its native packet x_2 to N_1 via a relay N_3.

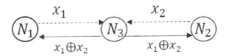

FIGURE 1.1: XOR network coding

In the first stage, both nodes send their native packets to the relay node separately. Between the received packets x_1 and x_2, an XOR coding is performed at the relay node. Then, instead of sending two packets one after the other, N_3 sends the coded packet $x_1 \oplus x_2$ all at once. Finally, the coded packet can be decoded with the help of its native packet at a destination node. The packet x_2, for example, can be decoded by doing $(x_1 \oplus x_2) \oplus x_1$ at N_1.

Because of its ease in both encoding and decoding procedures, XOR coding is gaining popularity. COPE [60], for example, uses XOR coding in its practical network coding system for wireless mesh networks. When a packet arrives at a node in COPE, it can be decoded. A router accomplishes this by encoding packets with the help of MAC layer feedback information. The network's throughput can be increased by using COPE.

This book focuses on the benefits of XOR-based network coding for broadcasting without feedback information in terms of reliability.

1.2.2 LINEAR NETWORK CODING

In multicast networks, linear network coding has been shown to be capable of achieving capacity limits from the source to the sink node [4]. The capacity limit, defined by the max-flow min-cut bound [61], is a graph theory consequence. A source node can deliver its information to the sink node at a largest rate of the min-cut between them.

In a multicast paradigm, where a source multicasts to numerous recipients using linear network coding technique. The source N_1 sends information to the sink nodes N_6 and N_7, as depicted in Fig. 1.2.

Each connection is given a d-dimensional column vector, with d being the maximum value of the max-flow of every non-source node. These vectors' entries are

Introduction

chosen from a finite field, such as the (Galois field) $GF(q)$, where q is any positive integer. $d = 2$ in the example presented in Fig. 1.2. Furthermore, the chosen finite field is $GF(2)$. $[1\ 0]^T$ is allocated to N_1N_2, where N_1N_2 denotes the channel from N_1 to N_2.

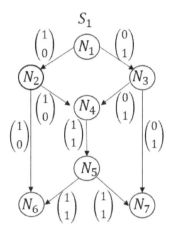

FIGURE 1.2: Illustration of the column vector assignment to different channels.

Furthermore, the vector assigned to a node's outgoing link is a linear combination of the equivalent vectors assigned to the incoming links. The vector assigned to N_4N_5, as shown in Fig. 1.2, is $[1\ 1]^T$, which is a summation of vectors allocated to the incoming links of N_4, i.e., N_2N_4 and N_3N_4.

Furthermore, the data to be sent is encoded as a d-dimensional row vector. The data flow on a channel is represented as the matrix product of the channel's allocated column vector and the information row vector. The data sent on a node's outgoing channel is a linear combination of the data delivered on the node's incoming channels in this scenario. Assume the data conveyed by N_1 is $[b_1\ b_2]$ in Fig. 1.2. The data transmitted on channel N_1N_2 is then represented by $[b_1\ b_2]\ [1\ 0]^T$, which is b_1. The data transferred at N_4 is represented by $[b_1\ b_2]\ [1\ 1]^T$, which is $b_1 + b_2$.

Finally, using the channel vectors [4], the information about the source can be obtained from received packets at a destination node.

Following the lead of [4], [62] solves the challenge of determining encoding functions at intermediate nodes, which are utilized to identify the vectors assigned to each channel. However, because linear solutions in any field may not exist for some networks, linear network coding is not always a sufficient solution [63].

1.2.2.1 Random linear network coding

Random linear network coding is presented in [5] to apply linear network coding to networks with unknown or changeable topologies. Unlike linear network coding, which assigns a deterministic vector to each link, random linear network coding allows nodes to choose a coefficient for each incoming packet at random across a finite field.

Consider the network topology shown in Fig. 1.2. The message transmitted on channel N_1N_2 can be represented by $\xi_1 b_1 + \xi_2 b_2$ when random linear network coding is used, where $\xi_1, \xi_2 \in GF(q)$, and q is an arbitrary positive integer. In reality, the same expression can be used to represent the message sent across each channel. Because the outcome of arithmetic operations on two values from a finite field falls into the same field in linear algebra, this is the case. The transmitting message on channel N_4N_5, for example, is $\xi_5(\xi_1 b_1 + \xi_2 b_2) + \xi_6(\xi_3 b_1 + \xi_4 b_2)$, which can be represented by $\xi_7 b_1 + \xi_8 b_2$, where $\xi_1, \xi_2, \xi_3, \xi_4; \xi_5, \xi_6, \xi_7, \xi_8 \in GF(q)$. When a source transmits $x_1, x_2, \ldots, x_M$ using random linear network coding, the data transmitted over the channel will be generated by $\sum_{i=1}^{M} \xi_i x_i$, where $\xi_i \in GF(q)$.

The encoding functions that determine the channel coefficients must be sent to the destination nodes in order for them to retrieve the coded packets. When a sufficient number of linear independent coded packets are received, decoding can be done using Gaussian elimination at the receiver.

1.2.2.2 Distributed random linear network coding

The global knowledge of coding coefficients is necessary for decoding in both [4] and [5]. However, in practise, it can be difficult to achieve. A distributed network coding approach is presented to overcome this problem [64]. A data-aided transmission system is used in [64], in which each transmitted data packet flowing on the network's edge having a header giving the coefficients Ξ_i of the combined packets.

Network coding can be used in a decentralized fashion in networks due to the overhead. Furthermore, in [64], a generation tag is introduced to identify coded packets. This approach overcomes the problem of incoming and outgoing packets belonging to the same set of packets being synchronized.

These methods, on the other hand, not only add to the temporal complexity of network coding-based approaches, but also to the difficulty of putting them into practise.

1.2.3 BENEFITS MADE BY NETWORK CODING

In wireless communications, bandwidth and power are the key resource limitations [65]. By allowing diverse information flows to share these resources, network coding can enhance bandwidth and power efficiency [56]. Furthermore, network coding has been shown to reduce the average delay of receiving a group of packets. It has also been used as an error control approach for ensuring reliable transmission.

1.2.3.1 Bandwidth efficiency

The gain in capacity when network coding is applied to error-free multicast [1] is the first scenario where network coding demonstrates its benefit. Multiple information streams can be transmitted simultaneously thanks to the coding at the intermediate node, resulting in a significant bandwidth gain over networks using the transitional

Introduction

store and forwards approach. The bandwidth benefit of network coding has been extensively studied since its publication, and network coding has been integrated in several bandwidth-efficient transmission systems.

1.2.3.2 The basic idea

The following diagram depicts the concept underlying network coding's increased bandwidth efficiency. The packet transmissions on the butterfly network are shown in Fig. 1.3, where source N_1 (N_2) sends native packet x_1 (x_2) to sink node N_4 (N_6). XOR coding is the network coding algorithm used here. A store and forward scheme is also depicted for comparison.

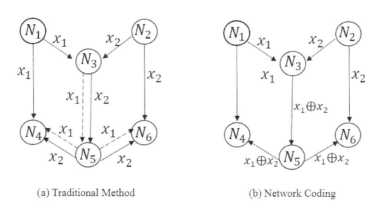

(a) Traditional Method (b) Network Coding

FIGURE 1.3: Traditional butterfly networks with routing or network coding schemes, where N_1 sends x_1 to N_4 and N_2 transmits x_2 to N_6

The network in Fig. 1.3(a) uses the classic store and forward technique. Only a single packet can be forwarded at a time by the central node N_3. As a result, N_3 must send two broadcasts to broadcast both x_1 and x_2 packets. As demonstrated in Fig. 1.3(b), network coding allows relay node N_3 to mix x_1 and x_2 and then broadcast $x_1 \oplus x_2$ in a single transmission. Nodes N_4 and N_6 can decode the coded packet based on their own packet. The number of transmissions can therefore be reduced. As a result, network coding improves bandwidth efficiency. In ad hoc networks [66–71] and wireless mesh networks [72], network coding has been used to boost bandwidth efficiency by lowering the number of transmissions.

Network coding is used in [71] to apply to one-to-all broadcast in multi-hop networks, where a node picks the neighboring nodes to forward the packet. The proposed techniques, which rely on two-hop topology information, are used to execute network coding. When compared to the non-coding strategy, the number of transmissions in networks using XOR based network coding methods can be reduced by up to 45 percent. The two-way relaying network, as shown in Fig. 1.1, is investigated in [67]. When network coding is used, the information exchange rate is computed. When compared to traditional methods that separate the processing of two unicast sessions, the exchange rate is found to be enhanced.

In a multicast model with n receivers, the average throughput has been investigated using either a linear network coding scheme or a random linear network coding method [73]. The average throughput is the average rate at which each individual receiver receives data. It is demonstrated using linear programming formulas that network coding provides a benefit in average throughput proportional to $\sqrt{n}$. Furthermore, for a large network, the average throughput with network coding more than doubles when compared to the throughput without network coding.

1.2.3.3 Other traffic configurations

The advantage of network coding in terms of bandwidth efficiency is not only limited to multicast [73] but also applies to other traffic arrangements, such as multiple unicast sessions [74].

COPE [60] is a technique, as described in Section 1.2.1, that improves the throughput of wireless multi-hop networks when traffic is sent in unicast form. [74] presents a theoretical throughput analysis improvement in multi-hop wireless networks using COPE-type network coding based on this. In addition, [74] proposes coding-aware routing and compares it to interference-aware routing, where coding-aware routing refers to selecting the path that maximizes coding opportunity. Finally, a linear programming optimization is used to determine the maximal unicast throughput. Compared with coding-oblivious routing algorithms, a route selection considering network coding opportunities leads to improved end-to-end throughput.

Routing [54, 75, 76] can enhance the performance of unicast sessions using netowrk coding even further. Random linear network coding is applied to supplement conventional routing methods in [75, 76]. It enables the source node to transmit random linear coded packets across numerous opportunistic pathways until the destination has a sufficient quantity of packets to decode. This protocol is best for transmitting files that are medium to large in size [75, 76]. When network coding is used for lossy wireless networks, [54] proposes a technique to optimize multipath routing and code rate to maximize throughput.

1.2.3.4 Undirected networks

In undirected networks, the benefits of throughput are investigated. The upper bound of throughput increase in a particular network topology, namely the combination network topology, is investigated in [77]. Combination networks are coded using linear network coding. It is proven that network coding can improve performance of combination networks. The improvement can be upper-bounded by a factor of $9/8$ by evaluating the cost of minimum multicast tree.

1.2.4 ENERGY EFFICIENCY

Energy efficiency has been demonstrated in networks employing network coding [55, 68, 78–80], where energy efficiency is often assessed by the energy consumed to transmit one bit of information.

Introduction

1.2.4.1 Multicast

The energy efficient routing in a traditional network, where a node just stores and forwards incoming data, is based on finding the minimum-energy multicast tree [81]. Some research has been done on calculating the minimum-energy multicast tree [82, 83], although this is usually NP-hard.

In [68], an alternate approach for low-energy multicast in a mobile ad hoc network is proposed. It takes a layered model of wireless networks and then generates a collection of realisable graphs, with each graph edge containing the energy per bit for the relevant channel. The edge-wise maximum of flow from the source to each sink node then characterizes the bit-rate demand on the edges. It is demonstrated that linear programming may be used to generate the lowest energy multicast energy-per-bit, which can be achieved by executing network coding but not routing.

1.2.4.2 Other traffic configurations

Ref. [78] investigates the power efficiency for a two-way relaying system using a network coding technique called a hybrid Automatic Repeat reQuest (ARQ) with incremental redundancy. It evaluates the average system power consumption and indicates that two-way relaying networks achieve high power efficiency by using network coding.

Refs. [55, 79, 80] examine the power efficiency for broadcasting in an all-to-all scenario. The network in [79] has a circular topology, whereas other network topologies, such as circular, rectangular grid, and random are investigated in [55]. Both linear network coding and random linear network coding techniques are used in ad-hoc networks in [79]. The simulation findings reveal that these coded networks have significantly higher energy efficiency than networks that use the classic store and forwards methodology.

In [55], techniques for improving power efficiency through network coding are presented. The goal of these algorithms is to reduce the number of transmissions. The total power is proportionate to the number of transmissions because it is expected that each transmission requires the same amount of energy. Furthermore, the power efficiency of various algorithms is assessed, demonstrating that network coding can enhance power efficiency by a constant factor in stationary networks and by a factor of $\log n$ in a dynamically changeable network with a number of n nodes [55].

1.2.5 DELAY PERFORMANCE

The adoption of network coding techniques does not necessarily result in a reduction in delay, because the encoding and decoding procedures, which require the collection of enough packets to proceed, can cause delays. Throughput efficiency and decoding delay are trade-offs: the higher the throughput, the more packets must be coded together, resulting in longer delay because the sink node must amass sufficient coded packets for decoding a single packet [84].

1.2.5.1 The average delay

In the case of big file transmissions, network coding can be employed to lessen the latency. The system considered in [58] is a single-hop one-to-all model, in which the source sends files through time-varying lossy channels to numerous recipients. The delay performance is represented as a closed-form expression. It is demonstrated that by increasing the system parameters, network coding can provide an arbitrary gain in delay. After examining several network coding methods, it was shown that random linear network coding has the shortest mean completion time, i.e., the time it takes for all receivers to get the entire file. In the case of packet streaming, however, these findings are not possible.

The mean time to finish the transmission of a number of packets from a source node to each recipient in lossy networks is investigated in [85]. The block of data is encoded using random linear network coding, and the receiver can recover the information once a sufficient number of coded packets have been gathered. As a result, instead of keeping track of which packets have been received, it simply record the demanded number of packets to complete decoding. This demanded number of packets is contained in the feedback data that is transmitted on a regular basis. The transmission process is described using a Markov chain, with the state representing the number of packets necessary for decoding at each recipient. In [86], the all-to-all model using random linear network coding is examined. In the initial transmission, the data block to be encoded is ready. Of contrast, the encoding in the systems suggested in this book is done on dynamically received packets.

The relationship between the average latency and several elements such as scheduling and the size of the coding buffer is investigated in [87]. One transmitter sends a data stream to a group of one-hop receivers, according to the system model. Furthermore, a network coding technique is presented, with the goal of lowering delay even more by transmitting non-coded packets. [88] derives the upper and lower bounds on the delay performance for multicast downlink transmissions. The anticipated delay is also taken into account. The average time it takes to receive a block of packets that allows decoding is known as the anticipated delay. At the transmitting node, the expected encoding and transmission delays are also investigated. It is demonstrated that a network can attain the lower bound on the predicted delay using the network coding strategy proposed in [87].

1.2.5.2 Delay distribution

Some delay sensitive applications require both the average delay and the worst-case decoding latency, which may be calculated using the whole probability distribution of delay. Note that the distribution of delay probability is studied in [59]. Delay distribution refers to the probability of successfully decoding all packets at a given delay. Ref. [59] takes into consideration an individual model in which a single source transmits randomly encoded linear network coding packets through erasure channels to all other nodes. Before the first broadcast, the broadcast message is available. The Markov chain is used to analyse a network with just three nodes and four nodes are proposed using a brute force method. In comparison, the message to be broadcast at a source node in the model in this book depends on the packets received by the source node that vary in time. Moreover, all broadcast on networks with arbitrary numbers of nodes is considered in this book.

[89] examines the exact probability of linearly independent N packages between K received in a receiver. The transmitted packets are assumed to be encoded by random linear network code and coefficients to be selected from an arbitrary finite file. The number of packets received is commensurate with the delay. The methods presented in this book can therefore be used to measure the distribution of delays.

1.2.6 RELIABILITY

Another significant advantage of network coding is the increased network reliability. In a non-coded wireless network, reliable communications usually require several retransmissions of the same information. To reduce the amount of retransmissions while keeping network reliability, a lot of research has been done. In lossy networks, network coding has lately been used as an error control mechanism for packet transmissions.

1.2.6.1 Retransmissions and network coding

To improve the reliability of packet transfers to numerous receivers, network coding is paired with retransmission mechanisms. Lossy networks use XOR-based network coding to limit the amount of retransmissions.

Many locations may have discontinuous lost packets when packets are broadcast in lossy wireless networks. Rather than transmitting each lost packet individually, the sender might encode the lost packets of different recipients and broadcast the coded packet to all recipients. Then, using the information of previously received packets, the lost packets can be recovered at a single receiver. Multiple receivers may recover their lost packets in a single transmission in this instance, reducing the overall number of retransmissions. An example is shown in Fig. 1.4, where node N_1 uses a relay N_2 to send packets x, y, and z to nodes N_3, N_4, and N_5. After broadcasting, N_3 did not receive y, both N_4 and N_5 lost packet z. In this case, N_2, the intermediate node, can network encode y and z and retransmit, allowing all sink nodes to retrieve their lost packets in a single retransmission if network coding is used.

In [57, 90, 91], the reliability of networks using network coding is investigated, and a reception report mechanism is used to inform the sender of each receiver's lost packets. In this case, you can generate the encoded packets with each receiver knowing of lost packets.

[90] applies network coding aided ARQ to access point (AP) based networks. In particular, a source node uses network code to transmit a selected combination of packets of the various recipients received unsuccessfully. All users are listening to all the packets in [90], and intended users can use overheard packets to decode the packet. This can reduce the number of transmissions.

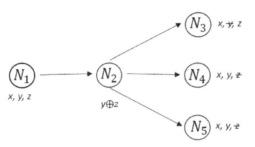

FIGURE 1.4: A network with five nodes using a combined network coding and retransmission scheme.

Ref. [57] proposes packet retransmission methods based on network coding for wireless broadcast in a one-to-all architecture. To encode the lost packets, XOR based network coding is used. Different retransmission schemes have been compared, it is determined that the scheme using dynamic network coding has the fewest retransmissions.

The following work in [91] focuses on determining the best coding set for lost packets in order to reduce the number of retransmissions. A coloring-based heuristic algorithm can be used to generate this coding set. The encoding of lost packets is based on a packet-loss table, which contains information about each receiver's packet loss. Every receiver must provide feedback in order to keep the packet-loss table up to date. Feedback, on the other hand, is costly in terms of bandwidth and energy efficiency in wireless communication, particularly in broadcast scenarios. As a result, [92–94] propose retransmission schemes without feedback information.

The authors of [92, 93] propose a two-stage broadcast scheme in which each node broadcasts its native packet in the first stage and an XOR coded packet in the second. In the second stage, they look into the best number of packets to encode in order to reduce the expected number of retransmissions given the connection probability of each channel. However, in terms of the expected number of retransmissions, this coding method does not always outperform non-coded networks. In this book, we develop several network coding schemes allowing a node to opt out of coding when it has a negative impact, so that it can achieve at least the performance of non-coded networks.

Ref. [94] considers a one-to-all model in which a single transmitter sends multiple streams over lossy channels to multiple destinations. A retransmission method based on XOR is presented, with the goal of providing a fair service to all customers in terms of service time and goodput. The transmitter in the proposed scheme estimates the reception status of all receivers without incurring any additional costs. After examining all feasible coding sets, the scheduler then selects the frames that provide the highest performance within fairness requirements. The XOR operation is used to encode multiple selected frames. The effectiveness of this retransmission scheme is demonstrated in a real-world setting.

Introduction

Furthermore, the reliability gain is quantified in [53,95], which compares network coding to traditional error control protocols like ARQ and Forward Error Correction (FEC). The systems under consideration have tree topologies with an equal number of children in each multicast tree. The number of retransmissions expected by the source node under various error control protocols is calculated. It is hypothesized that the reliability gain achieved by network coding increases logarithmically corresponding to the number of receivers in a multicasting group when compared to a scheme with simplified ARQ, based on numerical comparison. In the following work [53], this hypothesis is proven.

1.2.6.2 Combination of routing and network coding

Combining routing with network coding [96,97], in which an individual node mixes different received packets heading to the same destination [98], is another way to improve reliability. Fig. 1.5 shows an example.

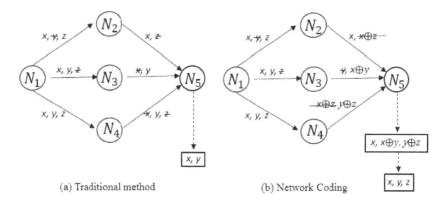

FIGURE 1.5: Routing vs Network Coding.

Assume that node N_1 wants to deliver packets x, y, and z to node N_5. The traditional way for these three packets delivery is shown in Fig. 1.5(a). For this example, N_5 cannot receive packet z. While for the case with combined network coding and multipath routing as illustrated in Fig. 1.5(b). N_5 can receives packets x, $x \oplus y$, and $y \oplus z$, which can be used to decode all three packets x, y, and z. It can be seen that network coding can improve packet transmission reliability with proper coding design.

The reliability performance for mesh networks with poor channel can be enhanced by using opportunistic routing [54,76]. Any node can overhear the transmission in traditional opportunistic routing. As a result, multiple nodes may hear the same packet and forwards it unnecessarily. This problem, on the other hand, can be solved using network coding. A node, for example, generates a random linear combination of the received packets in MORE [76]. Multiple nodes can avoid repetition by broadcasting coded packets with different coding coefficients. As a result, a destination can receive the packets correctly with high probability.

1.3 NETWORK CODING DESIGN CHALLENGES

When design network coding schemes for wireless networks, some challenges we need to solve.

For the XOR-based coding schemes, the advantage is the low coding and decoding complexity [59], however, their decodability of coded packets should be considered. An incorrect XOR coding design can have a negative impact on a network. A decoding may be unable to process in some cases due to a lack of native packets. Without a native packet, packets x_1, x_2, and x_3, for example, cannot be decoded from coded packets $x_1 \oplus x_2$, $x_2 \oplus x_3$, and $x_1 \oplus x_3$. Furthermore, because enough packets must be obtained before coding can begin, the number of native packets to be encoded has an impact on the encoding and decoding delay.

Several strict conditions must be met before linear network coding can be used in a network. First and foremost, the network's channels must be fixed and error-free. Second, each channel has a vector assigned to it, and the vector for a node's outgoing channel is a linear combination of the vectors assigned to its incoming channels. Finally, these vectors and network topology must be known centrally. Real channels, on the other hand, have errors and a constantly changing topology. As a result, liner network coding is primarily used for theoretical analysis and is rarely used in real-world networks.

The decentralized random linear network coding schemes [5] can work in real-time channels with errors. However, it is important to design parameters with each system in mind. One of the important parameters is the number of packets to be combined into a coded packet. It has an impact on the system performance in terms of throughput and decoding delay. More specifically, the higher the throughput and the longer the delay, the more packets are combined into a coded packet. As a result, more packets should be combined into one coded packet to achieve higher throughput. If some packets are delay sensitive, however, fewer packets should be encoded into a single coded packet. Another important parameter is the size of the finite field for the coefficients. It has an impact on the likelihood of decoding and the overhead length. As the finite field size increases, the greater the probability that the encoded packets are linearly independent, making the packets easier to decode. The overhead and bandwidth required to transmit these coefficients, however, are higher.

1.4 ORGANIZATION OF THE BOOK

We look at multiple sources communicating with multiple destinations via a common relay in Chapter 2. In the investigated system, we use nested codes to achieve multiple interpretations at different receivers. In addition, to maximize system capacity, an opportunistic scheduling (OS) technique is used at the relay. The studied system model combines the advantages of nested codes and operating systems. First, we go over the scheme's detailed coding process. The average channel capacity and outage probability for the nested network coded multiple interpretation system are then investigated. The upper bounds on the bit error probability with and without OS are also derived. Finally, we look into good codes for the investigated system and run simulations to verify the theoretical findings.

We look at network coded modulation schemes for multiple access relay channels in Chapter 3. The performance of the studied systems using distributed convolutional codes and network coded modulation is discussed. With Maximum Likelihood Sequence Detection (MLSD), an analytical upper bound on bit error probability performance for the investigated distributed systems is derived. For increasing channel Signal-to-Noise Ratio (SNR) values, the constructed bounds for the network coded modulation systems are shown to be asymptotically tight.

We focus on the implementation of multiple interpretations in multi-way relay channels (MWRC) with fading in Chapter 4, where multiple sources communicate with each other via a relay. Over the finite field, we first propose a novel nested convolutional lattice codes (NCLC) that can achieve the MI for each source in two time slots. The NCLC's codeword error rate (WER) is then given a theoretical upper bound. We improve our NCLC even more by creating a code design criterion that minimizes the derived WER. Based on our code design criterion, we construct a specific NCLC in simulations. The results of simulations show that our code can achieve multiple interpretations for each source in two time slots, confirming the derived upper bound in the high normalized signal-to-effective-noise ratio region.

We propose a new nested non-binary LDGM code with lattice in Chapter 5 to achieve multiple interpretations with low-complexity decoding methods. We design these novel codes in a multi-access wireless relay network and deduce the detailed coding process at both the relay and destination nodes. We develop a corresponding complexity reduced Lattice-based Extended Min-Sum (L-EMS) decoding method at the destination node. In terms of significant complexity reduction, the performance of the low complexity decoding algorithm is quite good. In addition, we propose a Lattice-based Monte Carlo method for obtaining good non-binary nested LDGM codes. In addition, it has been widely demonstrated in the literature that a lattice-based coding system can approach the Shannon capacity limit.

In Chapter 6, we propose an approach based on estimate-and-forward (EF) applied network-coded two-way relay channels to forwards soft information at the relay with a rate of 1 b/sample. We use different delivery schemes based on a preset threshold because the probability distribution of soft information varies depending on the signal-to-noise ratio (SNR) of the source-to-relay channel. We apply joint trellis coded quantization/modulation (TCQ/M) to the network-coded EF protocol in the low SNR region. We use an alternative scheme of decode-and-forward (DF) or EF with a refined TCQ/M codebook in the high SNR region. Over fading channels, both schemes outperform the amplify-and-forward (AF) and DF schemes, according to simulations.

We look at the reliability of lossy wireless networks in Chapter 7. Due to a variety of issues, such as channel fading, interference, or device mobility, wireless communications between devices can be lossy. Wireless communications unreliability can be random in some scenarios, making it easier to characterize from a stochastic standpoint. In light of this, lossy wireless networks, in which transmission between each pair of nodes is successful with a certain probability, have recently been investigated. In these networks, network coding can be used to improve the reliability of wireless communications. The reliability of a neighbor network coding scheme is investigated analytically in this chapter, where reliability is measured by the probability that every node in the network receives packets from every other node. It is demonstrated that the proposed neighbor coding scheme can improve network reliability. Also presented are closed-form upper and lower bounds on network reliability. In addition, an optimal neighbor coding scheme is discussed, which maximizes the likelihood that a packet broadcast from a designated source node will be received by all nodes in the network.

We investigate reliability performance for lossy broadcast networks in Chapter 8, in which each node in the network broadcasts to all the other nodes. The reliability is the ability to successfully decode the packets received from other nodes. For these networks, we develop a random neighbor network coding (RNNC) scheme to improve the reliability of the network. The RNNC scheme can adapt the lossy network channel conditions to generate network coded packets based on received packets. Both theoretical analysis and simulations demonstrate that the proposed scheme can significantly improve network reliability.

1. R. Ahlswede, N. Cai, S. Y. R. Li, and R. W. Yeung, "Network Information Flow," *IEEE Trans. Inform. Theory*, vol. 46, no. 4, pp. 1204–1216, July 2000.

2. R. W. Yeung and N. Cai, *Network error correction, part i: Basic concepts and upper bounds,* Communications in Information and Systems, vol. 6, no. 1, pp. 19–36, 2006.

3. R. W. Yeung, and N. Cai, Network error correction, part ii: lower bounds, Communications in Information and Systems, pp. 37–54, 2006.

4. S. Li, R. Yeung, and N. Cai, "Linear network coding," *IEEE/ACM Trans. Netw.*, vol. 49, no. 2, pp. 371–381, Feb. 2003.

5. T. Ho, M. Medard, R. Koetter, D. Karger, M. Effros, M. Jun and B. Leong, "A random linear network coding approach to multicast," *IEEE Transactions on Information Theory*, vol. 52, no. 10, pp. 4413–4460, 2006.

6. N. Cai and R. W. Yeung, *Network coding and error correction*, Proceedings of the IEEE Information Theory Workshop, 2002, pp. 119–122.

7. M. Di Renzo, M. Iezzi, and F. Graziosi, *Beyond routing via network coding: An overview of fundamental information-theoretic results*, Proceedings of IEEE 21st International Symposium on Personal Indoor and Mobile Radio Communications (PIMRC), 2010, pp. 2745–2750.

8. H. Balli, X. Yan, and Z. Zhang, *On randomized linear network codes and their error correction capabilities*, IEEE Transactions on Information Theory, vol. 55, no. 7, pp. 3148–3160, 2009.

9. R. Koetter and F. R. Kschischang, *Coding for errors and erasures in random network coding*, IEEE Transactions on Information Theory, vol. 54, no. 8, pp. 3579–3591, 2008.

10. H. Mahdavifar and A. Vardy, *List-decoding of subspace codes and rank-metric codes up to singleton bound,* in IEEE International Symposium on Information Theory Proceedings (ISIT), 2012, pp. 1488–1492.

11. L. Song, R. W. Yeung, and N. Cai, *A separation theorem for single-source network coding,* IEEE Transactions on Information Theory, vol. 52, no. 5, pp. 1861–1871, 2006.

12. Z. Li and B. Li, *Network coding in undirected networks,* in the 38th Annu. Conf. Information Sciences and Systems, 2004.

13. P. Sanders, S. Egner, and L. Tolhuizen, *Polynomial time algorithms for network information flow,* in 15th ACM Symposium on Parallel Algorithms and Architectures, 2003.

14. K. Jain, V. V. Vazirani, R. Yeung, and G. Yuval, *On the capacity of multiple unicast sessions in undirected graphs,* in International Symposium on Information Theory, 2005, pp. 563–567.

15. Z. Li, B. Li, J. Dan, and L. L. Chi, *On achieving optimal throughput with network coding,* in Proceedings of 24th Annual Joint Conference of the IEEE Computer and Communications Societies, vol. 3, 2005, pp. 2184–2194 vol. 3.

16. Z. Li, B. Li, and L. L. Chi, *A constant bound on throughput improvement of multicast network coding in undirected networks,* IEEE Transactions on Information Theory, vol. 55, no. 3, pp. 1016–1026, 2009.

17. S. Zhang, *Hot topic: physical-layer network coding,* in Proceeding of ACM Mobicom, 2006.

18. B. Nazer and M. Gastpar, *Reliable physical layer network coding,* Proceedings of the IEEE, vol. 99, no. 3, pp. 438–460, 2011.

19. Y. E. Sagduyu, D. Guo and R. Berry, *On the delay and throughput of digital and analog network coding for wireless broadcast,* in 42nd Annual Conference on Information Sciences and Systems, 2008, pp. 534–539.

20. R. H. Y. Louie, Y. Li, and B. Vucetic, *Practical physical layer network coding for two way relay channels: performance analysis and comparison,* IEEE Transactions on Wireless Communications, vol. 9, no. 2, pp. 764–777, 2010.

21. M. Kim, C. W. Ahn, M. Medard, and M. Eros, *On minimizing network coding resources: An evolutionary approach,* in Proceedings of NetCod, 2006.

22. H. Yao and E. Verbin, *Network coding is highly non-approximable,* in 47th Annual Allerton Conference on Communication, Control, and Computing, 2009, pp. 209–213.

23. C. K. Ngai and R. W. Yeung, *Network coding gain of combination networks,* in IEEE Information Theory Workshop, 2004, pp. 283–287.

24. M. Xiao, M. Medard, and T. Aulin, *A binary coding approach for combination networks and general erasure networks,* in IEEE International Symposium on Information Theory, 2007, pp. 786–790.

25. B. Nazer and M. Gastpar, "Compute-and-forward: Harnessing interference through structured codes," IEEE Trans. Inf. Theory, vol. 57, no. 10, pp. 6463–6486, Oct. 2011.

26. U. Erez and R. Zamir, "Achieving 1/2 log (1+ SNR) on the AWGN channel with lattice encoding and decoding," IEEE Trans. Inf. Theory, vol. 50, no. 10, pp. 2293-2314, Oct. 2004.

27. C. Feng, D. Silva, and F. R. Kschischang, "An algebraic approach to physicallayer network coding," IEEE Trans. Inf. Theory, vol. 59, no. 11, pp. 7576–7596, Nov. 2013.

28. S. Zhang, S. C. Liew, and P. P. Lam, "Hot topic: Physical-layer network coding," in Proc. ACM MobiCom, 2006, pp. 358–365.

29. L. Ong, S. J. Johnson, and C. M. Kellett, "An optimal coding strategy for the binary multi-way relay channel," IEEE Commun. Lett., vol. 14, no. 4, pp. 330–332, Apr. 2010.

30. L. Ong, C. Kellett, and S. Johnson, "Capacity theorems for the AWGN multiway relay channel," in Proc. IEEE ISIT, Jun. 2010, pp. 664–668.

31. D. Gunduz, A. Yener, A. Goldsmith, and H. V. Poor, The multiway relay channel," IEEE Trans. Inf. Theory, vol. 59, no. 1, pp. 51–63, Jan. 2013.

32. J. Li, J. Yuan, R. Malaney, M. H. Azmi, and M. Xiao, "Network coded LDPC code design for a multi-source relaying system," IEEE Trans. Wireless Commun., vol. 10, no. 5, pp. 1538-1551, May 2011.

33. Y. Wu, P. A. Chou, and S. Y. Kung, "Information exchange in wireless networks with network coding and physical-layer broadcast," in Technical Report MSR-TR-2004-78, Microsoft Research, Redmond, WA, Aug. 2004.

34. S. Zhang and S. C. Liew, "Channel coding and decoding in a relay system operated with physical-layer network coding," IEEE J. Sel. Areas Commun., vol. 27, no. 5, pp. 788–796, Jun. 2009.

35. S. Zhang, S. C. Liew, and P. Lam, "Physical layer network coding," in Proc. ACM MOBICOM, Los Angeles, 2006.

36. B. Nazer and M. Gastpar, "Reliable physical layer network coding," Proc. IEEE, vol. 99, no. 3, pp. 438–460, 2011.

37. J. N. Laneman, D. N. C. Tse, and G. W. Wornell, "Cooperative diversity in wireless networks: Efficient protocols and outage behavior," IEEE Trans. Inf. Theory, vol. 50, no. 12, pp. 3062–3080, Dec. 2004.

38. C. Deqiang and J. N. Laneman, "Modulation and demodulation for cooperative diversity in wireless systems," IEEE Trans. Wireless Commun., vol. 5, no. 7, pp. 1785–1794, 2006.

39. Y. Li, B. Vucetic, T. Wong, and M. Dohler, "Distributed turbo coding with soft information relaying in multihop relay networks," IEEE J. Sel. Areas Commun., vol. 24, no. 11, pp. 2040–2050, Nov. 2006.

40. K. S. Gomadam and S. A. Jafar, "Optimal relay functionality for snr maximization in memoryless relay networks," IEEE J. Sel. Areas Commun., vol. 25, no. 2, pp. 390–401, Feb. 2007.

41. J. Li, M. Karim, J. Yuan, Z. Chen, Z. Lin, and B. Vucetic, "Novel soft information forwarding protocols in two-way relay channels," IEEE Trans. Veh. Technol., vol. 62, no. 5, pp. 2374–2381, Jun. 2013.

42. M. A. Karim, J. Yuan, Z. Chen, and J. Li, "Soft information relaying in fading channels," IEEE Wireless Commun. Lett., vol. 1, no. 3, pp. 233–236, Jun. 2012.

43. J. Li, J. Yuan, R. Malancy, M. Xiao, and W. Chen, "Full-diversity binary frame-wise network coding for multiple-source multiple-relay networks over slow-fading channels," IEEE Trans. Veh. Technol., vol. 61, no. 3, pp. 1346–1360, Mar. 2012.

44. M. Xiao, J. Kliewer, and M. Skoglund, "Design of network codes for multiple-user multiple-relay wireless networks," IEEE Trans. Commun., vol. 60, no. 12, pp. 3755–3766, Dec. 2012.

45. J. Barros and S. D. Servetto, "Network information flow with correlated sources," IEEE Trans. Inf. Theory, vol. 52, no. 1, pp. 155–170, 2006.

46. L. Ong, R. Timo, and S. Johnson, "The finite field multi-way relay channel with correlated sources: Beyond three users," in 2012 IEEE International Symposium on Information Theory Proceedings (ISIT), 2012, pp. 791–795.

47. M. Karim, T. Yang, J. Yuan, Z. Chen, and I. Land, "A novel soft forwarding technique for memoryless relay channels based on symbol-wise mutual information," IEEE Commun. Lett., vol. 14, no. 10, pp. 927–929, 2010.

48. M. Karim, J. Yuan, Z. Chen, and J. Li, "Analysis of mutual information based soft forwarding relays in awgn channels," in IEEE Global Telecommunications Conference (GLOBECOM 2011), 2011, pp. 1–5.

49. C. Hausl and J. Hagenauer, "Iterative network and channel decoding for the two-way relay channel," in 2006 IEEE International Conference on Communications (ICC), vol. 4, 2006, pp. 1568–1573.

50. G. Wang, W. Xiang, and J. Yuan, "Multihop compute-and-forward for generalized two-way relay channels," Transactions on emerging telecommunications technologies, pp. 1–13, In Press.

51. S. Zhang, Y. Zhu, and S. C. Liew, "Soft network coding in wireless two-way relay channels," J. Communication and Networks, Special Issues on Network Coding, vol. 10, no. 4, 2008.

52. C. Tao, T. Ho, and J. Kliewer, "Memoryless relay strategies for two-way relay channels," IEEE Trans. Commun., vol. 57, no. 10, pp. 3132–3143, 2009.

53. M. Ghaderi, D. Towsley, and J. Kurose, "Reliability gain of network coding in lossy wireless networks," in Proceedings of IEEE INFOCOM, 2008, pp. 2171–2179.

54. X. Zhang and B. Li, "Optimized multipath network coding in lossy wireless networks," IEEE Journal on Selected Areas in Communications, vol. 27, no. 5, pp. 622–634, 2009.

55. C. Fragouli, J. Widmer, and J. Y. Le Boudec, "Ecient broadcasting using network coding," IEEE/ACM Transactions on Networking, vol. 16, no. 2, pp. 450–463, 2008.

56. C. Fragouli, J.-Y. L. Boudec, J.Widmer, "Network coding: an instant primer," SIGCOMM Comput. Commun. Rev., vol. 36, no. 1, pp. 63–68, 2006.

57. D. Nguyen, T. Tran, T. Nguyen, and B. Bose, "Wireless broadcast using network coding," IEEE Transactions on Vehicular Technology, vol. 58, no. 2, pp. 914–925, 2009.

58. A. Eryilmaz, A. Ozdaglar, M. Medard, and E. Ahmed, "On the delay and throughput gains of coding in unreliable networks," IEEE Transactions on Information Theory, vol. 54, no. 12, pp. 5511–5524, 2008.

59. M. Nistor, D. E. Lucani, T. T. V. Vinhoza, R. A. Costa, and J. Barros, "On the delay distribution of random linear network coding," IEEE Journal on Selected Areas in Communications, vol. 29, no. 5, pp. 1084–1093, 2011.

60. S. Katti, H. Rahul, H. Wenjun, D. Katabi, M. Medard, and J. Crowcroft, *Xors in the air: Practical wireless network coding,* IEEE/ACM Transactions on Networking, vol. 16, no. 3, pp. 497–510, 2008.

61. P. Elias, A. Feinstein, and C.E. Shannon, *A note on the maximum flow through a network,* IRE Transactions on Information Theory, vol. 2, no. 4, pp. 117–119, 1956.

62. R. Koetter and M. Medard, *An algebraic approach to network coding,* IEEE/ACM Transactions on Networking, vol. 11, no. 5, pp. 782–795, 2003.

63. R. Dougherty, C. Freiling, and K. Zeger, *Insuciency of linear coding in network information flow,* IEEE Transactions on Information Theory, vol. 51, no. 8, pp. 2745–2759, 2005.

64. P. A. Chou, Y. Wu, and K. Jain, *Practical network coding,* in Allerton Conf. commun. Control, Computing, 2003.

65. J. Liu, D. Goeckel, and D. Towsley, *Bounds on the gain of network coding and broadcasting in wireless networks,* in Proceedings of the 26th IEEE International Conference on Computer Communications, 2007, pp. 724–732.

66. T. Ho, R. Koetter, M. Medard, D. R. Karger, and M. Eros, *The benefits of coding over routing in a randomized setting,* in Proceedings of IEEE International Symposium on Information Theory, 2003, p. 442.

67. Y. Wu, P. A. Chou, and S.-Y. Kung, *Information exchange in wireless networks with network coding and physical-layer broadcast,* in Conference on Information Sciences and Systems, 2005.

68. Y. Wu, P. A. Chou, and S.-Y. Kung, *Minimum-energy multicast in mobile ad hoc networks using network coding,* IEEE Transactions on Communications, vol. 53, no. 11, pp. 1906–1918, 2005.

Introduction

69. S. Deb, M. Eros, T. Ho, D. R. Karger, R. Koetter, D. S. Lun, M. Medard, and N. Ratnakar, *Network coding for wireless applications: A brief tutorial,* in Proceedings of Wireless Ad-hoc Sensor Workshop, 2005.

70. J. Liu, D. Goeckel, and D. Towsley, *The throughput order of ad hoc networks employing network coding and broadcasting,* in IEEE Military Communications Conference (MILCOM), 2006, pp. 1–7.

71. L. Li, R. Ramjee, M. Buddhikot, and S. Miller, *Network coding-based broadcast in mobile ad-hoc networks,* in Proceedings of IEEE 26th International Conference on Computer Communications, 2007, pp. 1739–1747.

72. R. Prasad, H. Wu, D. Perkins, and N.-F. Tzeng, *Local topology assisted xor coding in wireless mesh networks,* in 28th International Conference on Distributed Computing Systems Workshops, 2008, pp. 156–161.

73. C. Chekuri, C. Fragouli, and E. Soljanin, *On average throughput and alphabet size in network coding,* IEEE Transactions on Information Theory, vol. 52, no. 6, pp. 2410–2424, 2006.

74. S. Sengupta, S. Rayanchu, and S. Banerjee, *An analysis of wireless network coding for unicast sessions: The case for coding-aware routing,* in Proceedings of IEEE 26th International Conference on Computer Communications, 2007, pp. 1028–1036.

75. S. Chachulski, M. Jennings, S. Katti, and D. Katabi, *Trading structure for randomness in wireless opportunistic routing,* in Proceedings of the 2007 conference on Applications, technologies, architectures, and protocols for computer communications, Kyoto, Japan, 2007.

76. ——, *Trading structure for randomness in wireless opportunistic routing,* SIGCOMM Comput. Commun. Rev., vol. 37, pp. 169–180, 2007.

77. S. Maheshwar, Z. Li, and B. Li, *Bounding the coding advantage of combination network coding in undirected networks,* IEEE Transactions on Information Theory, vol. 58, no. 2, pp. 570–584, 2012.

78. J. Choi and D. To, *Energy eciency of harq-ir for two-way relay systems with network coding,* Preceeding of 18th European Wireless Conference, pp. 1–5, 2012.

79. D. Platz, D. H. Woldegebreal, and H. Karl, *Random network coding in wireless sensor networks: Energy eciency via cross-layer approach,* in Proceeding of IEEE 10th International Symposium on Spread Spectrum Techniques and Applications, 2008, pp. 654–660.

80. C. Fragouli, J. Widmer, and J. Y. Le Boudec, *On the benefits of network coding for wireless applications,* in Preceeding of 4th International Symposium on Modeling and Optimization in Mobile, Ad Hoc and Wireless Networks, 2006, pp. 1–6.

81. J. E. Wieselthier, G. D. Nguyen, and A. Ephremides, *On the construction of energye cient broadcast and multicast trees in wireless networks,* in Proceedings of IEEE Nineteenth Annual Joint Conference of the IEEE Computer and Communications Societies., vol. 2, 2000, pp. 585–594 vol. 2.

82. D. S. Lun, N. Ratnakar, R. Koetter, M. Medard, E. Ahmed, and L. Hyunjoo, *Achieving minimum-cost multicast: a decentralized approach based on network coding,* in 24th Annual Joint Conference of the Computer and Communications Societies, vol. 3, 2005, pp. 1607–1617 vol. 3.

83. X. Yufang and E. M. Yeh, *Distributed algorithms for minimum cost multicast with network coding,* IEEE/ACM Transactions on Networking, vol. 18, no. 2, pp. 379–392, 2010.

84. X. Li, C.-C. Wang, and X. Lin, *Throughput and delay analysis on uncoded and coded wireless broadcast with hard deadline constraints,* in Proceedings of IEEE INFOCOM, 2010, pp. 1–5.

85. D. E. Lucani, M. Medard, and M. Stojanovic, *Broadcasting in time-division duplexing: A random linear network coding approach,* in Workshop on Network Coding, Theory, and Applications, 2009, pp. 62–67.

86. ——, *On coding for delay - network coding for time-division duplexing,* IEEE Transactions on Information Theory, vol. 58, no. 4, pp. 2330–2348, 2012.

87. W.-L.. Yeow, A.T. Hoang, and C.-K. Tham, *Minimizing delay for multicast-streaming in wireless networks with network coding,* in Proceedings of IEEE INFOCOM, 2009, pp. 190–198.

88. ——, *On average packet delay bounds and loss rates of network-coded multicasts over wireless downlinks,* in IEEE International Conference on Communications, 2009, pp. 1–6.

89. O. Trullols-Cruces, J. M. Barcelo-Ordinas, and M. Fiore, *Exact decoding probability under random linear network coding,* IEEE Communications Letters, vol. 15, no. 1, pp. 67–69, 2011.

90. P. Larsson and N. Johansson, *Multi-user arq,* in Proceedings of IEEE Vehicular Technology Conference, vol. 4, 2006, pp. 2052–2057.

91. W. Fang, F. Liu, Z. Liu, L. Shu, and S. Nishio, *Reliable broadcast transmission in wireless networks based on network coding,* in IEEE Conference on Computer Communications Workshops (INFOCOM WKSHPS), 2011, pp. 555–559.

92. Z. Wang, M. Hassan, and T. Moors, *Efficient loss recovery using network coding in vehicular safety communication,* in IEEE Wireless Communications and Networking Conference (WCNC), 2010, pp. 1–6.

93. Z. Wang and M. Hassan, *Blind xor: Low-overhead loss recovery for vehicular safety communications,* IEEE Transactions on Vehicular Technology, vol. 61, no. 1, pp. 35–45, 2012.

94. F.-C. Kuo, K. Tan, X. Li, J. Zhang, and X. Fu, *Xor rescue: Exploiting network coding in lossy wireless networks,* in Proceedings of IEEE 6th Annual Communications Society Conference on Sensor, Mesh and Ad Hoc Communications and Networks, 2009, pp. 1–9.

95. M. Ghaderi, D. Towsley, and J. Kurose, *Network coding performance for reliable multicast,* in Proceedings of IEEE Military Communications Conference, 2007, pp. 1–7.

96. Z. Guo, B. Wang, P. Xie, W. Zeng, and J.-H. Cui, *Ecient error recovery with network coding in underwater sensor networks,* Ad Hoc Networks, vol. 7, no. 4, pp. 791–802, 2009.

97. Z. Yang, M. Li, and W. Lou, *R-code: Network coding-based reliable broadcast in wireless mesh networks,* Ad Hoc Networks, vol. 9, no. 5, pp. 788–798, 2011.

98. M. Medard and A. Sprintson, *Harnessing network coding in wireless systems,* in Network coding: Fundamentals and Applications. Boston: Elsevier/Academic Press, 2012.

2 Wireless Network Coded Systems for Multiple Interpretations

2.1 INTRODUCTION

Wireless network coding is an extension of network coding for wired networks to wireless packet networks. It incorporates wireless communication properties such as wireless channel propagation, omnidirectional transmission, and so on into the network coding concept. In comparison to wired network coding, the interaction between the Medium Access Control (MAC) and network coding becomes critical due to the wireless communication properties. Joint scheduling and wireless network coding are investigated in [1]. In that work, the entire network is partitioned into some conflict-free disjoint subnetworks, each with a minimum cost assignment. The network throughput is then optimized by using joint scheduling and network coding. Ref [2] investigates opportunistic scheduling for wireless network coding. The basic idea is to use opportunistic scheduling to dynamically change the network coding group size in order to maximize average throughput. The proposed wireless network coding scheme, on the other hand, is based on full side information, which means that each destination node must be aware of all packets transmitted by all source nodes except the one destined for this destination node. This assumption may be unrealistic in practise. In this case, the traditional solution would be to concatenate the packets received at the relay node and broadcast them to all destination nodes. As a result, the side information at the destination nodes cannot be used during decoding.

The authors present the concept of nested codes with multiple interpretations in [3]. The basic concept is that packets received at the relay node are encoded with a low code rate before being broadcast to the destination nodes. The whole codeword set is divided into many subsets, and the side information is used to identify the subset. If no side information is available, decoding will be performed using the entire set of codewords. As a result, depending on the available side information at each destination, decoding for codewords with different coding rates can be performed.

We consider Opportunistic Scheduling (OS) for wireless nested network coding, as well as transmit power optimisation and rate adaptation at the relay node in this chapter. By nested network coding, we mean that before transmission, information packets from each source node are encoded using the nested coding concept, and the received codewords are XORed at the relay node before broadcasting to the destination nodes.

DOI: 10.1201/9781003203803-2

The chapter outline is provided below. Section 2.2 describes the system model. Section 2.3 formulates the optimisation problem. Section 2.4 provides a performance analysis in terms of average channel capacity and outage probability. Section 2.5 proposes a network coded system based on nested codes. Section 2.6 introduces analytical bounds on error performance. Section 2.7 contains the results of the Code Search. The numerical and simulation results are presented in Section 2.8. Finally, in Section 2.9, conclusions are drawn.

2.2 SYSTEM MODEL

Consider a wireless network with a single relay node and a collection of source and destination nodes. The relay node relays data packets from the source nodes to the destination nodes. Let us denote the set of source nodes by $\mathfrak{S}$ and the set of destination nodes by $\mathfrak{T}$. A connection is established between a source node S and a destination node T via the relay node R, where $S \in \mathfrak{S}$ and $T \in \mathfrak{T}$. A directed hypergraph $\mathfrak{G} = (\mathfrak{N}, \mathfrak{E})$ $\mathfrak{G} = (\mathfrak{N}, \mathfrak{E})$ can be used to model the system, where $\mathfrak{N}$ is the set of nodes, $\mathfrak{N} = \mathfrak{S} \bigcup \mathfrak{T} \bigcup r$ is a set of hyperedges, and $\mathfrak{E}$ is a set of hyperedges. A hyperedge $(n, \mathfrak{D})$ is made up of the directed edges between a start node n and a set of end nodes $\mathfrak{D}$, where $n \in \mathfrak{N}$, $\mathfrak{D} \subset \mathfrak{N}$ and $\mathfrak{D}$ is non-empty. A $(n, \mathfrak{D})$ hyperedge represents broadcasting links between node n and the nodes in $\mathfrak{D}$.

This scenario is illustrated in Fig. 2.1, in which four source nodes transmit information packets to four destination nodes via the relay node. In this case, $\mathfrak{S} = \{a, b, c, d, f\}$ and $\mathfrak{T} = \{b, c, e, f\}$ are used. The packets originated from the source nodes a, c, d, and f are denoted by i_a, i_c, i_d, i_f, respectively. We assume that all source nodes have packets to send and that all packets have the same κ bit length. The packets received by the relay node are linearly combined over a finite field GF(2) and then broadcast to all destination nodes in the network coding concept. The destination nodes can successfully decode the packets that are intended for them if all of the destination nodes have the side information from the neighboring source nodes. If the destination node b, for example, is aware of the packets transmitted by source nodes c, a, and f, it can successfully decode the packet i_d. However, in real-world systems, complete side information from source nodes may not be available. Node b, for example, may only have side information from neighbors a and c; in this case, node b is unable to decode the packet i_d.

We consider a system in which information blocks, such as i_a, are encoded at the source node a before being transmitted to the relay node in this paper. At the relay node, the received codewords from all source nodes are linearly combined using the XOR operation before being broadcast to all destination nodes. Assume that transmission from the source nodes to the relay node is error-free. The relay node then generates a codeword.

$$\mathbf{C} = i_a \mathbf{G}_a \oplus i_c \mathbf{G}_c \oplus i_d \mathbf{G}_d \oplus i_f \mathbf{G}_f, \qquad (2.1)$$

where $\mathbf{G}_a$, $\mathbf{G}_c$, $\mathbf{G}_d$, $\mathbf{G}_f$ are the generator matrices for the corresponding source nodes with a $\kappa/\eta N_T$, and N_T is the cardinality of the set of destination nodes $\mathfrak{T}$.

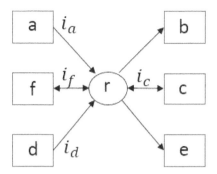

FIGURE 2.1: A network coding group with four source nodes and four destination nodes and one relay node.

With its own side information, each destination node decodes the received codeword. Let Θ_ω represent the destination node's side information. After demodulation, ω, $\omega \in \mathfrak{T}$ and $\hat{\mathbf{C}}$ are the received codewords. The received codeword is combined with the side information at node ω.

$$\tilde{\mathbf{C}}_\omega = \hat{\mathbf{C}} \oplus \Theta_\omega. \tag{2.2}$$

$\tilde{\mathbf{C}}_\omega$ denotes the received mixed codewords which are unknown to the node ω. It represents a corrupted version of the codeword with a code rate $(N_T - |\Theta_\omega|)\kappa/(\eta N_T)$, where $|\Theta_\omega|$ is the length of the set of the destination nodes which are known by node ω. The receiver at node ω can then decode $\tilde{\mathbf{C}}_\omega$ with the corresponding code rate. Therefore, a single codeword $\mathbf{C}$ can be interpreted differently by different destination nodes. This is also the basic idea of nested codes. For more detailed description of the nested codes, please refer to [3].

With the network coding scheme described above, each destination node can interpret the received message based on its own side information. However, because a codeword is broadcast by the relay node, the relay node must choose a suitable data rate to ensure its reliable transmission. The transmission data rate should adapt to the channel with the worst link condition for a reliable transmission, that is, to ensure that every destination node receives the codeword broadcasted by the relay node reliably. However, for channels with good link conditions within the broadcasting group, this rate may be too conservative. In [2,6,7], this problem has been extensively researched.

In this chapter, we look at wireless nested network coding with opportunistic scheduling and consider simultaneous power and rate adaptation.

2.3 OPTIMIZATION FORMULATION

Given a wireless ad hoc network which can be described as a hypergraph $\mathfrak{G}$ with source nodes $\mathfrak{S}$, destination nodes $\mathfrak{T}$ and a relay node r, the hyperedge $(r, \mathfrak{T})$ defines the broadcast link between r and all the destination nodes. Let γ_i, $i \in \{1, \cdots, N_T\}$, be the instantaneous channel Signal to Noise Ratio (SNR) for the channel between the relay node and the ith destination node. For a reliable transmission, the broadcasting data rate should adapt to the channel with the worst link condition. According to the Shannon capacity theorem [4, 5], the instantaneous maximum achievable data rate for such a channel is equal to $C = \log_2(1 + \gamma_{min})$, where

$$\gamma_{min} = \min_{\forall i, i \in \{1,2,\cdots,N_T\}} \gamma_i. \quad (2.3)$$

The overall data rate for the set of destination nodes $\mathfrak{T}$ is then $N_T C$.

Define a subgraph of $\mathfrak{G}$ as $\mathfrak{G}_e = (\mathfrak{N}_e, \mathfrak{E}_e)$ where $\mathfrak{N}_e \subset \mathfrak{N}$ and $\mathfrak{N}_e = \mathfrak{S}_e \bigcup \mathfrak{T}_e \bigcup r$, $\mathfrak{S}_e \subset \mathfrak{S}$ and $\mathfrak{T}_e \subset \mathfrak{T}$. For each subgraph $\mathfrak{G}_e$, there is a corresponding maximum broadcasting data transmission rate C_e. Let P_{av} be the average transmit power at the relay node, $f_\Gamma(\gamma)$ be the Probability Density Function (PDF) of the instantaneous received SNR denoted by Γ^1. Suppose the instantaneous transmit power $P_t(\gamma)$ can be adapted to the received instantaneous SNR Γ subject to an average power constraint P_{av}. Denote by N_s the cardinality of the set $\mathfrak{T}_e$. Then the overall rate of $\mathfrak{T}_e$ is $N_s C_e$. Let $\mathfrak{B}$ be the set of all subgraphs of $\mathfrak{G}$. Then the optimization problem can be formulated as follows:

$$\max_{\mathfrak{G}_e \in \mathfrak{B}} (N_s C_e)$$

$$s.t.: \int_\gamma P_t(\gamma) f_\Gamma(\gamma) d\gamma \leq P_{av},$$

$$\mathfrak{G}_e \in \mathfrak{B}, \mathfrak{G}_e \subseteq \mathfrak{G}, \mathfrak{N}_e \subseteq \mathfrak{N},$$

$$\mathfrak{S}_e \subseteq \mathfrak{S}, \mathfrak{T}_e \subseteq \mathfrak{T}. \quad (2.4)$$

One implementation of the above optimization problem is the opportunistic scheduling algorithm [2]. With opportunistic scheduling, the received instantaneous SNR are ordered at the relay node. The first order statistic is the minimum of the received SNR. The subset $\mathfrak{T}_e$ of the destination nodes is partitioned by the ordered received instantaneous SNR. Let γ_k be the kth order received SNR. The subset $\mathfrak{T}_e$ with γ_k as the smallest received instantaneous SNR then contains $N_s = N_T - k + 1$ destination nodes with $C_e = \log_2(1 + \gamma_k)$. At the relay node, only the packets which are destined to the destination nodes in $\mathfrak{T}_e$ are linearly combined by the XOR operation. The XORed codeword is broadcasted by the relay node. Each destination node of $\mathfrak{T}_e$ receives a corrupted codeword with channel corruption and decodes it with its own side information.

---
[1] In the following, throughout this chapter, we denote by an upper case letter a random variable and by the corresponding lower case letter its realization.

2.4 ANALYSIS OF THE AVERAGE CHANNEL CAPACITY

For a time varying channel with channel side information at both the transmitter and the receiver side, the average channel capacity, given an average power constraint, can be expressed as [8]

$$C = \int_{\gamma_0}^{\infty} \log_2(\frac{\gamma}{\gamma_0}) f_\Gamma(\gamma) d\gamma, \qquad (2.5)$$

where $f_\Gamma(\gamma)$ is the PDF of the instantaneous channel SNRm denoted by Γ, and γ_0 is the cutoff SNR value which must satisfy

$$\int_{\gamma_0}^{\infty} (\frac{1}{\gamma_0} - \frac{1}{\gamma}) f_\Gamma(\gamma) d\gamma = 1. \qquad (2.6)$$

At a time instant i, if the instantaneous channel SNR, which is assumed to be perfectly known at both the transmitter and receiver side, is below the cutoff value γ_0, then no data is transmitted. Otherwise, the transmitter power and transmission rate will adapt to the instantaneous channel SNR to maximize the channel capacity.

With optimal simultaneous power and rate adaptation to the channel condition of the worst connection between the relay node and the destination nodes in $\mathfrak{T}_e$, the total average capacity of the hyperedge $(n, \mathfrak{T}_e)$ can be expressed as

$$C_{av} = N_s \int_{\gamma_0}^{\infty} \log_2(\frac{\gamma_k}{\gamma_0}) f_{\Gamma_k}(\gamma_k) d\gamma_k. \qquad (2.7)$$

Here γ_k is supposed to be the minimum SNR value of the subgraph $\mathfrak{G}_e$. Assuming that the channel between the relay node and each destination node is subjected to the Rayleigh fading, then the PDF of the channel SNR is given by [9]

$$f_\Gamma(\gamma) = \frac{1}{\bar{\gamma}} e^{-\gamma/\bar{\gamma}}, \quad \gamma \geq 0 \qquad (2.8)$$

where $\bar{\gamma}$ is the average instantaneous channel SNR.

The corresponding Cumulative Distribution Function (CDF) is then

$$F_\Gamma(\gamma) = \int_0^\gamma f_\Gamma(t) dt = 1 - e^{-\gamma/\bar{\gamma}}. \qquad (2.9)$$

By using the order statistics [10], the PDF of the kth order instantaneous SNR value γ_k can be expressed as

$$f_{\Gamma_k}(\gamma_k) = N_T \binom{N_T - 1}{k - 1} F_\Gamma(\gamma)^{k-1} [1 - F_\Gamma(\gamma)]^{N_T - k} f_\Gamma(\gamma), \qquad (2.10)$$

where

$$\binom{N_T - 1}{k - 1} = \frac{(N_T - 1)!}{(N_T - k)!(k - 1)!} \qquad (2.11)$$

is the binomial coefficient. By inserting (2.10) into (2.7), we get

$$C_{av} = N_s \int_{\gamma_0}^{\infty} \log_2\left(\frac{\gamma_k}{\gamma_0}\right) N_T \binom{N_T - 1}{k - 1}$$
$$\cdot \frac{1}{\bar{\gamma}} \left(1 - e^{-\gamma_k/\bar{\gamma}}\right)^{k-1} e^{-(N_T-k+1)\gamma_k/\bar{\gamma}} d\gamma_k. \qquad (2.12)$$

The exponential expansion gives

$$C_{av} = N_s N_T \binom{N_T - 1}{k - 1} \frac{1}{\bar{\gamma}} \sum_{i=0}^{k-1} \frac{(k-1)!}{i!(k-1-i)!}$$
$$\cdot (-1)^i \int_{\gamma_0}^{\infty} \log_2\left(\frac{\gamma_k}{\gamma_0}\right) e^{-(N_T-k+i+1)\gamma_k/\bar{\gamma}} d\gamma_k. \qquad (2.13)$$

Let $\mu(i) = (N_T - k + i + 1)\gamma_0/\bar{\gamma}$. Then we get

$$C_{av} = N_s N_T \binom{N_T - 1}{k - 1} \frac{\gamma_0}{\bar{\gamma}} \sum_{i=0}^{k-1} \frac{(k-1)!}{i!(k-1-i)!}$$
$$\cdot (-1)^i \log_2(e) \int_1^{\infty} \ln(t) e^{-\mu(i)t} dt. \qquad (2.14)$$

Denote by $E_m(\alpha)$ the exponential integral of order m defined as

$$E_m(\alpha) = \int_1^{+\infty} t^{-m} e^{-\alpha t} dt, \quad \alpha \geq 0. \qquad (2.15)$$

Eq. (2.14) can then be written as

$$C_{av} = N_s N_T \binom{N_T - 1}{k - 1} \frac{\gamma_0}{\bar{\gamma}} \sum_{i=0}^{k-1} \frac{(k-1)!}{i!(k-1-i)!}$$
$$\cdot (-1)^i \log_2(e) \frac{E_1(\mu(i))}{\mu(i)}. \qquad (2.16)$$

Now let us look at γ_0, since γ_0 must satisfy

$$\int_{\gamma_0}^{\infty} \left(\frac{1}{\gamma_0} - \frac{1}{\gamma_k}\right) f_{\Gamma_k}(\gamma_k) d\gamma_k = 1, \qquad (2.17)$$

so

$$N_T \binom{N_T - 1}{k - 1} \sum_{i=0}^{k-1} \frac{(k-1)!}{i!(k-1-i)!} (-1)^i \int_1^{\infty} \left(1 - \frac{1}{t}\right) e^{-\mu(i)t} dt = \bar{\gamma}. \qquad (2.18)$$

Since
$$\int_1^\infty \left(1 - \frac{1}{t}\right) e^{-\mu(i)t} dt = E_0(\mu(i)) - E_1(\mu(i)) \tag{2.19}$$

and
$$E_0(t) = e^{-t}/t, \tag{2.20}$$

we get,
$$\sum_{i=0}^{k-1} \frac{N_T!}{(N_T-k)!i!(k-1-i)!}(-1)^i \left[\frac{e^{-\mu(i)}}{\mu(i)} - E_1(\mu(i))\right] = \bar{\gamma}. \tag{2.21}$$

γ_0 can then be determined by solving (2.21). Note that γ_0 is inside $\mu(i)$.

The outage probability which corresponds to the situation when $\gamma < \gamma_0$, i.e., no transmission is allowed, is given by

$$\begin{aligned}
P_{out} &= \int_0^{\gamma_0} f_{\Gamma_k}(\gamma_k) d\gamma_k = F_{\Gamma_k}(\gamma_0) \\
&= \sum_{j=k}^{N_T} \binom{N_T}{j} F_\Gamma(\gamma_0)^j (1 - F_\Gamma(\gamma_0))^{N_T-j} \\
&= \sum_{j=k}^{N_T} \binom{N_T}{j} (1 - e^{-\gamma_0/\bar{\gamma}})^j e^{-(N_T-j)\gamma_0/\bar{\gamma}}.
\end{aligned} \tag{2.22}$$

When $k = 1$, i.e., $\mathfrak{G}_e = \mathfrak{G}$, (2.21) becomes

$$N_T \left[\frac{e^{-\mu(0)}}{\mu(0)} - E_1(\mu(0))\right] = \bar{\gamma}, \tag{2.23}$$

where
$$\mu(0) = N_T \frac{\gamma_0}{\bar{\gamma}}. \tag{2.24}$$

It can be seen as $\bar{\gamma} \to +\infty$, γ_0 tends to one. Therefore, γ_0 should be in the interval $[0, 1]$. For this case, (2.16) becomes

$$\begin{aligned}
C_{av} &= N_s N_T \frac{\gamma_0}{\bar{\gamma}} \log_2(e) \frac{E_1(\mu(0))}{\mu(0)} \\
&= \log_2(e) E_1(N_T \frac{\gamma_0}{\bar{\gamma}}) \\
&= \log_2(e) \left[\frac{e^{-\mu(0)}}{\mu(0)} - \frac{\bar{\gamma}}{N_T}\right] \\
&= \frac{\log_2(e)}{N_T} \left[\frac{\bar{\gamma}}{\gamma_0} e^{-N_T \frac{\gamma_0}{\bar{\gamma}}} - \bar{\gamma}\right].
\end{aligned} \tag{2.25}$$

For channel side information at the receiver only, i.e., the power at the transmitter is constant, the average channel capacity can be expressed as [2, 11, 12]

$$\begin{aligned}
C_{av} &= N_s \int_0^\infty \log_2(1+\gamma_k) f_{\Gamma_k}(\gamma_k) d\gamma_k \\
&= N_s N_T \binom{N_T-1}{k-1} \frac{1}{\bar{\gamma}} \cdot \sum_{i=0}^{k-1} \frac{(k-1)!}{i!(k-1-i)!}(-1)^i \\
&\quad \cdot \int_0^\infty \log_2(1+\gamma_k) e^{-(N_T-k+i+1)\gamma_k/\bar{\gamma}} d\gamma_k \\
&= N_s \log_2(e) N_T \binom{N_T-1}{k-1} \frac{1}{\bar{\gamma}} \cdot \sum_{i=0}^{k-1} \frac{(k-1)!}{i!(k-1-i)!}(-1)^i e^{v(i)} \frac{E_1(v(i))}{v(i)},
\end{aligned}$$
(2.26)

where $v(i) = (N_T - k + i + 1)/\bar{\gamma}$. For the case of $k = 1$,

$$C_{av} = \log_2(e) N_s e^{N_T/\bar{\gamma}} E_1(N_T/\bar{\gamma})$$

$$= \log_2(e) e^{N_T/\bar{\gamma}} \cdot \left(-E - \ln(N_T/\bar{\gamma}) + \sum_{j=1}^{+\infty} \frac{(-1)^{j+1}(N_T/\bar{\gamma})^j}{j \cdot j!} \right)$$
(2.27)

where E is the Euler constant ($E = 0.577215665$) and

$$E_1(x) = -E - \ln x + \sum_{j=1}^{\infty} \frac{(-1)^{j+1} x^j}{j \cdot j!}.$$
(2.28)

2.5 NETWORK CODED SYSTEM BASED ON NESTED CODES

We introduced a nested coded system based on rate compatible convolutional codes (RCPC) and opportunistic scheduling in this section (OS). The system model is depicted in Fig. 2.2. It is divided into three major sections: nested code encoding at source nodes, OS and RCPC code implementation at the relay node, and multiple interpretations at destination nodes.

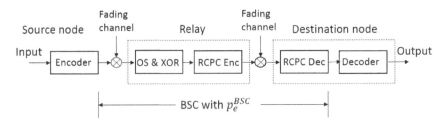

FIGURE 2.2: The coding process of a nest network coded system

In the investigated scheme, each source node's transmitted information is encoded with a different (linearly independent) generator and then forwarded to the relay node. Because the transmitted data will be nested at the relay node via the XOR operation, the rate of nested packets will be easily too high to decode if the source data is encoded with a high rate convolutional code. For example, if we XOR different source information with the same encode rate R', the nested code rate will be $|\mathfrak{S}| \cdot R'$ ($|\mathfrak{S}|$ denotes the number of source nodes), which is easily higher than one. However, such a high rate can only be decoded at the sink nodes [13] with perfect prior knowledge.

Alternatively, we select a low rate $R = R'/|\mathfrak{S}|$ code to encode the information at each source node, and then perform the XOR operation at the relay node, mathematically,

$$\begin{aligned} c_{nested} &= i_1 G_1 \oplus i_2 G_2 \oplus \cdots \oplus i_{|\mathfrak{S}|} G_{|\mathfrak{S}|} \\ &= [i_1, i_2, \ldots, i_{|\mathfrak{S}|}][G_1, G_2, \ldots, G_{|\mathfrak{S}|}]^T, \end{aligned} \quad (2.29)$$

where $G_1, G_2, \ldots, G_{|\mathfrak{S}|}$ are mutually linearly independent generators. $\oplus$ denotes the XOR operation.

We assume that all of the links in our model experience slow Rayleigh fading on their own. The instantaneous signal-to-noise ratio (SNR) of the link between the source node s and the relay node r at time instant μ is given by [2]

$$\gamma_{sr}(\mu) = \frac{P \mid h_{sr}(\mu) \mid^2}{\sigma_{sr}^2}, \quad (2.30)$$

where P is the constant transmission power, $h_{sr}(\mu)$ is the channel coefficient of the link between s and r at the time instant μ, σ_{sr}^2 denotes the variance of the additive white Gaussian noise at the relay node r.

We assume that the relay node has perfect channel state information (CSI). The relay node can then sort the instantaneous SNRs of the links between the source and relay nodes into an ascending array. Let γ_k denote the kth ordered received instantaneous SNR. Here, parameter k is called the scheduling level for the network coding group [15]. N_d denotes the cardinality of the set of destination nodes $\mathfrak{D}$, i.e., $N_d = |\mathfrak{D}|$. For the scheduling level k, the size of destination group selected by OS is $N_t = N_d - k + 1$, where each destination node has the instantaneous SNR higher than or equal to the kth smallest instantaneous SNR. In this case, the instantaneous capacity per unit bandwidth can be expressed as

$$C_{inst}^k = N_t \log_2(1 + \gamma_k). \quad (2.31)$$

At the relay node, only the packets destined to the selected destination nodes are linearly combined by the XOR operation. Let $\mathfrak{S}_e$, $\mathfrak{S}_e \subseteq \mathfrak{S}$, denote the set of source nodes whose packets are selected to be forwarded to destination nodes. We can get a new codeword with OS at relay

$$c_{nested}^{OS} = \bigoplus_{j \in \mathfrak{S}_e} i_j G_j \quad (2.32)$$
$$= [i_1, i_2, \ldots i_j, \ldots, i_{|\mathfrak{S}_e|}][G_1, G_2, \ldots G_j, \ldots, G_{|\mathfrak{S}_e|}]^T.$$

To adapt the data rate maximized by OS, a class of RCPC codes is employed at the relay node. In this work, a rate $1/3$, constraint length 3 optimum distance spectrum (ODS) code [5 7 7] is used as the parent code. The corresponding puncturing matrices of the rate $1/3$, $2/5$, $2/4$, and $2/3$ are chosen as [14].

When the channel condition is good enough, the maximum capacity can be efficiently utilized by first selecting the highest puncturing rate $2/3$. If the received instantaneous SNR value is less than the predetermined threshold, the complementary information at that position will be resent.

2.5.1 SOFT-DECISION DECODING WITH NESTED CODES

At different destination nodes, after the RCPC decoder, we get an output $L_{c_{nested}^{OS}}(i)$, the log-likelihood ratio (LLR) of the i-th bit in c_{nested}^{OS}. The codeword c_{nested}^{OS} can be decomposed as

$$c_{nested}^{OS} = \underbrace{\bigoplus_{l \notin \kappa_d} i_l G_l}_{c_u} \oplus \underbrace{\bigoplus_{l' \in \kappa_d} i_{l'} G_{l'}}_{c_c}, \quad (2.33)$$

where κ_d denotes the indices of the information prior known to the destination node d. c_u represents the collection of unknown information. c_c is the collection of information known to the receiver d, which can therefore be canceled. The LLR of the i-th bit in c_u can be computed as

$$L_{c_u(i)} \triangleq \frac{\Pr[c_u(i) = 0]}{\Pr[c_u(i) = 1]}$$
$$= \begin{cases} L_{c_{nested}^{OS}}(i) = \log \frac{\Pr[c_u \oplus c_c(i) = 0]}{\Pr[c_u \oplus c_c(i) = 1]} & \text{if } c_c = 0 \\ -L_{c_{nested}^{OS}}(i) = \log \frac{\Pr[c_u \oplus c_c(i) = 1]}{\Pr[c_u \oplus c_c(i) = 0]} & \text{if } c_c = 1 \end{cases} \quad (2.34)$$

Because the above cancellation operation only changes the sign of the LLR, no information is lost. The calculated LLR L_{c_u} is then obtained, which is the estimated soft information of unknown packets to receiver d. We can separate all of the desired information at different destination nodes using the linearly independent feature of nested codes.

2.6 ANALYTICAL BOUNDS ON THE BIT ERROR PROBABILITY

The part between the output of the encoder at the source nodes and the input of the last decoder at the destination nodes can be regarded as a binary symmetric channel (BSC), with a crossover probability $\mathcal{P}_e^{\text{BSC}}$, as shown in Fig.2.2.

We begin by obtaining the expression for the crossover probability $\mathcal{P}_e^{\text{BSC}}$. As will be seen later, this probability will be used to calculate the bit error probability for the links from the source nodes to the destination nodes. Let $\mathcal{P}_c^{\text{XOR}}$ be the correct bit transmission probability for network coded packets (with XOR operation) at the relay, and $\mathcal{P}_c^{r \to d}$ be the correct bit transmission probability for the link after the RCPC decoder from the relay node r to the destination node d. The crossover probability of $\mathcal{P}_e^{\text{BSC}}$ can then be written as

$$\mathcal{P}_e^{\text{BSC}} \leq 1 - \mathcal{P}_c^{\text{XOR}} \mathcal{P}_c^{r \to d}. \tag{2.35}$$

Only selected packets from different source nodes will be network encoded in the case of OS at the relay. Because the likelihood of different packets suffering from bit errors at the same positions is extremely low, especially at high SNR, the correct bit transmission probability for network coded packets can be approximated as

$$\mathcal{P}_c^{\text{XOR}} \approx \prod_{s \in \mathfrak{S}_e} \mathcal{P}_c^{s \to r}, \tag{2.36}$$

where $\mathcal{P}_c^{s \to r}$ is the correct bit transmission probability for the link from s to r. For uncoded BPSK modulation over a Rayleigh fading channel, the correct bit transmission probability of $\mathcal{P}_c^{s \to r}$ can be expressed as [16]

$$\begin{aligned} \mathcal{P}_c^{s \to r} &= 1 - \mathcal{P}_e^{s \to r} \\ &\geq 1 - \left[\frac{1}{2} \left(1 - \sqrt{\frac{\overline{\gamma}_{sr} R_1}{1 + \overline{\gamma}_{sr} R_1}} \right) \right] \\ &= \frac{1}{2} \left(1 + \sqrt{\frac{\overline{\gamma}_{sr} R_1}{1 + \overline{\gamma}_{sr} R_1}} \right), \end{aligned} \tag{2.37}$$

where $\mathcal{P}_e^{s \to r}$ is the bit error probability for the link from s to r. $\overline{\gamma}_{sr} = |\overline{h}_{sr}|^2 E_b/N_0$ is the received average SNR, where E_b/N_0 denotes SNR per information bit. R_1 is the code rate of the convolutional encoder at source nodes.

Let $\mathcal{P}_e^{r \to d}$ be the bit error probability for the link from r to d. Then, the correct bit transmission probability $\mathcal{P}_c^{r \to d}$ can be expressed as

$$\mathcal{P}_c^{r \to d} = 1 - \mathcal{P}_e^{r \to d}. \tag{2.38}$$

With the RCPC encoder-decoder pair over the link from r to d, $\mathcal{P}_e^{r \to d}$ can be upper bounded by [17–19],

$$\mathcal{P}_e^{r \to d} \leq \frac{1}{p} \sum_{\mu=d_{free}^2}^{\infty} c_\mu P_\mu, \qquad (2.39)$$

where p is the puncturing period, d_{free}^2 is the free distance of the RCPC codes, c_μ is the total number of bit errors for different incorrect track paths at distance φ, and P_φ is the pairwise error probability. For BPSK modulation over an uncorrelated slow Rayleigh fading channel with perfect channel estimation and soft decision decoding at the receiver, P_φ can be written as

$$P_\mu = q^\mu \sum_{\delta=0}^{\mu-1} \binom{\mu-1+\delta}{\delta} (1-q)^\delta, \qquad (2.40)$$

$$q = \frac{1}{2}\left(1 - \sqrt{\frac{\overline{\gamma}_{rd} R_2}{1 + \overline{\gamma}_{rd} R_2}}\right), \qquad (2.41)$$

where R_2 is the code rate of the RCPC codes. Inserting (2.39) into (2.38), we can get a lower bound on $\mathcal{P}_c^{r \to d}$. Inserting this lower bound and (2.36) into (2.35), we then get an upper bound on the crossover probability of $\mathcal{P}_e^{BSC}$.

Subsequently, by treating the channel between a source node s and a destination node d as a BSC channel with the crossover probability of $\mathcal{P}_e^{BSC}$, we can then get an upper bound on the bit error probability for the convolutional coded link from the source node s to the destination node d [20],

$$\mathcal{P}_e^{s \to d} < \frac{1}{m} \sum_{\mu=d_{free}^1}^{\infty} a_\mu \left[4\mathcal{P}_e^{BSC}\left(1-\mathcal{P}_e^{BSC}\right)\right]^{\frac{\mu}{2}}. \qquad (2.42)$$

Here m is the number of message bits fed to the convolutional encoder at the source node s, d_{free}^1 is the free distance of the convolutional code, and a_μ is the number of paths at a distance μ from the all-zero path. Since the process of the cancelation at destination nodes only discards the known information from the received network coded packets, it will not change the amount of error and the packets' length. Therefore, the crossover probability for BSC is still $\mathcal{P}_e^{BSC}$ after the cancelation process at destination nodes.

Finally, the upper bound for the bit error probability of the proposed scheme with OS is given by Eq. (2.43), where

$$\mathcal{P}_e^{BSC} = \left\{1 - \left[\prod_{s \in \mathfrak{S}_c} \frac{1}{2}\left(1 + \sqrt{\frac{\overline{\gamma}_{sr} R_1}{1 + \overline{\gamma}_{sr} R_1}}\right)\right] \left\{1 - \frac{1}{p}\sum_{\mu=d_{free}^2}^{\infty}\left[c_\mu \left(\frac{1}{2} - \frac{1}{2}\sqrt{\frac{\overline{\gamma}_{rd} R_2}{1 + \overline{\gamma}_{rd} R_2}}\right)^\mu \right. \right. \right.$$

$$\left. \left. \left. \times \sum_{\delta=0}^{\mu-1} \binom{\mu-1+\delta}{\delta} \left(\frac{1}{2} + \frac{1}{2}\sqrt{\frac{\overline{\gamma}_{rd} R_2}{1 + \overline{\gamma}_{rd} R_2}}\right)^\delta\right]\right\}\right\}$$

$$(2.43)$$

Similarly, we can obtain the upper bound for the bit error probability of the scheme without OS by replacing $\mathfrak{G}_e$ with $\mathfrak{G}$.

2.7 CODE SEARCH

To achieve multiple interpretations with nested codes, the code design is assumed to satisfy the following criterion:

1. the generators assigned to different nodes should be mutually linearly independent.
2. the rate of "stacked" generator matrix should be less than 1.
3. the selected code should not be a catastrophic convolutional code.

We construct several good codes based on the modified FAST algorithm in [21], which are presented in Table 2.1. The generator matrices are given in octal form converting binary words into the corresponding octal words, for example, the generator polynomial (110101001) in binary notation would be represented as (6 5 1) in octal form. Then, we can choose different rows of one generator matrix as different linearly independent generators.

In this work, we choose a rate $4/6$ code from Table 2.1. For a network coded system with four source nodes, each node can be assigned to one generator polynomial with the code rate of $1/6$ encoder, for example, $G_1 = [5\ 6\ 5\ 6\ 7\ 4]$ for node 1, $G_2 = [7\ 0\ 7\ 3\ 6\ 2]$ for node 2, $G_3 = [4\ 5\ 2\ 6\ 5\ 0]$ for node 3 and $G_4 = [6\ 1\ 5\ 7\ 5\ 2]$ for node 4. Tables 2.2 and 2.3 illustrate the code performance for different code rates, which can form a rate $4/6$ code at the relay when assign the generate matrices to different nodes. For example, if the system contains 2 nodes, then we can assign any two rate $2/6$ codes to the two nodes or any one of the rate $1/6$ codes and any one

TABLE 2.1
Table of good codes

Rate	Memory number	Generator matrices	d_{free}
2/3	2	6 5 1 7 2 5	5
2/4	2	3 7 1 6 4 7 6 3	8
3/4	2	5 4 3 2 4 6 5 5 6 1 4 3	6
4/6	2	5 6 5 6 7 4 7 0 7 3 6 2 4 5 2 6 5 0 6 1 5 7 2 5	8

TABLE 2.2
Distance spectrum for code rates $1/6$ **and** $2/6$

Rate	Generator matrices	d_{free}	$(c_\mu, \mu = d_{free}, d_{free}+1, \ldots, d_{free}+6)$ $(a_\mu, \mu = d_{free}, d_{free}+1, \ldots, d_{free}+6)$
1/6	(5 6 5 6 7 4)	12	$(1, 0, 0, 0, 2, 0, 5)$ $(1, 0, 0, 0, 1, 0, 2)$
	(7 0 7 3 6 2)	10	$(2, 1, 0, 3, 0, 0, 8)$ $(1, 1, 0, 1, 0, 0, 2)$
	(4 5 2 6 5 0)	8	$(1, 0, 0, 0, 2, 0, 2)$ $(1, 0, 0, 0, 1, 0, 1)$
	(6 1 5 7 2 5)	11	$(1, 0, 0, 0, 0, 4, 0)$ $(1, 0, 0, 0, 0, 2, 0)$
2/6	$\begin{pmatrix} 5\ 6\ 5\ 6\ 7\ 4 \\ 7\ 0\ 7\ 3\ 6\ 2 \end{pmatrix}$	9	$(5, 18, 14, 19, 87, 200, 374)$ $(2, 4, 4, 5, 14, 26, 48)$
	$\begin{pmatrix} 5\ 6\ 5\ 6\ 7\ 4 \\ 4\ 5\ 2\ 6\ 5\ 0 \end{pmatrix}$	8	$(3, 0, 10, 0, 40, 0, 131)$ $(2, 0, 3, 0, 12, 0, 28)$
	$\begin{pmatrix} 5\ 6\ 5\ 6\ 7\ 4 \\ 6\ 1\ 5\ 7\ 2\ 5 \end{pmatrix}$	9	$(2, 0, 7, 13, 11, 40, 68)$ $(1, 0, 3, 4, 3, 19, 12)$
	$\begin{pmatrix} 7\ 0\ 7\ 3\ 6\ 2 \\ 4\ 5\ 2\ 6\ 5\ 0 \end{pmatrix}$	8	$(1, 2, 5, 9, 9, 24, 26)$ $(1, 1, 2, 4, 3, 7, 7)$
	$\begin{pmatrix} 7\ 0\ 7\ 3\ 6\ 2 \\ 6\ 1\ 5\ 7\ 2\ 5 \end{pmatrix}$	8	$(2, 0, 4, 16, 24, 24, 28)$ $(1, 0, 2, 6, 6, 6, 6)$
	$\begin{pmatrix} 4\ 5\ 2\ 6\ 5\ 0 \\ 6\ 1\ 5\ 7\ 2\ 5 \end{pmatrix}$	8	$(1, 2, 0, 9, 2, 10, 48)$ $(1, 1, 0, 4, 1, 3, 12)$

of the rate 3/6 codes. In Tables 2.2 and 2.3, d_{free} is the free distance of the code, c_μ and a_μ represent the sum of bit errors for error events of distance μ and the number of error paths with distance μ, respectively.

2.8 NUMERICAL AND SIMULATION RESULTS

2.8.1 AVERAGE CHANNEL CAPACITY AND OUTAGE PROBABILITY

In this section, we present some numerical and simulation results for the systems under consideration. The average channel capacity in bits per second per Hz (bits/s/Hz) and the outage probability were taken into account. The total average channel capacity optimization for nested wireless network coding with opportunistic

TABLE 2.3
Distance spectrum for code rates $3/6$ **and** $4/6$

Rate	Generator matrices	d_{free}	$(c_\mu, \mu = d_{free}, d_{free}+1, \cdots, d_{free}+6)$ $(a_\mu, \mu = d_{free}, d_{free}+1, \cdots, d_{free}+6)$
3/6	$\begin{pmatrix} 5 & 6 & 5 & 6 & 7 & 4 \\ 7 & 0 & 7 & 3 & 6 & 2 \\ 4 & 5 & 2 & 6 & 5 & 0 \end{pmatrix}$	8	$(8, 42, 95, 208, 580, 1744, 4971)$ $(3, 10, 18, 35, 88, 228, 570)$
	$\begin{pmatrix} 5 & 6 & 5 & 6 & 7 & 4 \\ 7 & 0 & 7 & 3 & 6 & 2 \\ 6 & 1 & 5 & 7 & 2 & 5 \end{pmatrix}$	8	$(2, 29, 102, 162, 349, 1218, 3622)$ $(1, 7, 19, 30, 55, 162, 429)$
	$\begin{pmatrix} 5 & 6 & 5 & 6 & 7 & 4 \\ 4 & 5 & 2 & 6 & 5 & 0 \\ 6 & 1 & 5 & 7 & 2 & 5 \end{pmatrix}$	8	$(10, 42, 75, 214, 527, 1612, 4644)$ $(3, 11, 13, 34, 76, 205, 545)$
	$\begin{pmatrix} 7 & 0 & 7 & 3 & 6 & 2 \\ 4 & 5 & 2 & 6 & 5 & 0 \\ 6 & 1 & 5 & 7 & 2 & 5 \end{pmatrix}$	8	$(9, 15, 24, 90, 285, 845, 2294)$ $(4, 4, 7, 23, 51, 135, 313)$
4/6	$\begin{pmatrix} 5 & 6 & 5 & 6 & 7 & 4 \\ 7 & 0 & 7 & 3 & 6 & 2 \\ 4 & 5 & 2 & 6 & 5 & 0 \\ 6 & 1 & 5 & 7 & 2 & 5 \end{pmatrix}$	8	$(76, 782, 2571, 11049, 49770, 211356, 917121)$ $(15, 97, 269, 973, 3841, 14646, 57026)$

scheduling and simultaneous power and rate adaption is considered. In comparison, fixed scheduling is investigated, in which the subgraph always contains $(N_T - k + 1)$ destination nodes with the smallest kth order instantaneous channel SNR.

Both simulation and theoretical results are provided for fixed k level scheduling. We only show the simulation results for opportunistic scheduling. For the simulation, we generate N_T instantaneous SNR values at each time instant at random and list them in ascending order. The total average channel capacity is then calculated for each SNR order. In the case of opportunistic scheduling, we compute the average channel capacity for various k level scheduling and then select the one with the highest channel capacity.

Figs. 2.3 and 2.4 plot average channel capacity versus average channel SNR for wireless network coding with $N_T = 5$ and $N_T = 10$ destination nodes, respectively. The simulation results for fixed scheduling (labeled as FS) agree very well with the theoretical analysis. When compared to a fixed power (FP) policy at the relay node, power adaptation (legend as AP) can improve total average channel capacity. For opportunistic scheduling (legend as OS), the gain with power adaptation is more

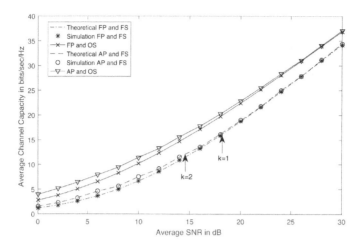

FIGURE 2.3: Opportunistic scheduling against fixed scheduling with $N_T = 5$, $k = 1, 2$ and rate adaptation with/without optimal power allocation.

significant; at 10 bits/s/Hz, the gain in terms of average SNR is about 2 dB for $N_T = 5$ and about 3 dB for $N_T = 10$.

It can also be seen that as the order of SNR value increases from 1 to 2, so does the average capacity. This is due to the fact that as k increases, the network coding group size decreases and the worst channel SNR improves. There is a tradeoff between the size of the network coding group and the worst channel link quality, as shown by (2.7). The smaller the network coding group becomes as the worst link quality improves. Opportunistic scheduling provides the best solution. That explains why, in Figs. 2.3 and 2.4, the opportunistic scheduling algorithm always outperforms the fixed scheduling algorithm.

The outage probability for the previously investigated wireless network coding scheme with power adaptation is depicted in Fig. 2.5. For fixed scheduling, the outage probability improves as the order statistic increases from $k = 1$ to 2. This is due to the fact that as the minimum SNR improves, the likelihood of this minimum SNR being less than the cutoff rate SNR decreases. The larger the wireless network coding group, the more likely it is that some users will experience deep fading. This means that a larger group has a higher outage probability, as illustrated in Fig. 2.5. When compared to fixed scheduling, opportunistic scheduling with power adaptation can significantly reduce the likelihood of an outage. Because opportunistic scheduling and power adaptation can significantly improve the network coding group's worst SNR, the probability of this worst SNR being less than the cutoff rate SNR becomes small. When the number of users available for scheduling increases, the improvement becomes more pronounced. This explains why the outage probability in Fig. 2.5 is lower for $N_T = 10$ than for $N_T = 5$.

Wireless Network Coded Systems for Multiple Interpretations 41

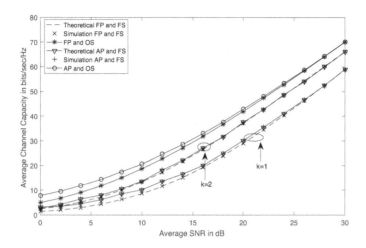

FIGURE 2.4: Opportunistic scheduling against fixed scheduling with $N_T = 10$, $k = 1, 2$ and rate adaptation with/without optimal power allocation.

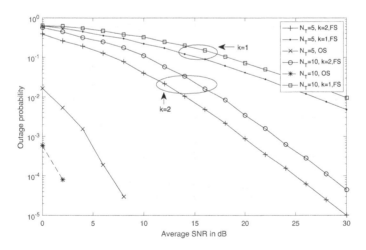

FIGURE 2.5: Outage probability for opportunistic and fixed scheduling for N = 5,10, k = 1,2 with optimal power and rate adaptation.

2.8.2 THE PERFORMANCE OF OS

We present computer simulation results for the performance of OS and nested codes in this section. We assume in the simulation results that all nodes have the same transmission power P and that all links experience the same independent Rayleigh fading. We introduce a fixed scheduling (FS) with the fixed level k to compare the performance of the system with and without OS. When the same RCPC codes are used, the system with OS is equivalent to the system without OS in the case of scheduling level k fixed to 1.

When the code rate of the RCPC codes is increased, the performance on the bit error probability deteriorates. Indeed, the impact of the worst code performance on the overall system is much lower than that of the uncoded Rayleigh fading channel between the sources and the relay node. As a result, we only investigate one code rate of the RCPC codes for analysing the bit error probability performance of the proposed scheme, because the performances on the bit error probability of the whole system for the different code rates of the RCPC codes are the same at high SNR.

We plot analytical upper bounds and simulation curves of the bit error probability under different OS selections in Fig. 2.6, assuming a constant RCPC code rate of $1/3$. The bit error probability performance under consideration here is that of the received XORed packets prior to the last decoder at destination nodes, namely $\mathcal{P}_e^{\text{BSC}}$ in Eq. 2.35. It has a low complexity and follows the same trend as $\mathcal{P}_e^{s \to d}$ in Eq. 2.42 when $\mathcal{P}_e^{\text{BSC}} < 1/2$. Furthermore, regardless of the sets of destination nodes to which the source nodes are destined, there are only four packets generated from four source

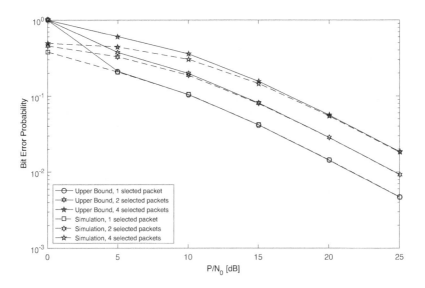

FIGURE 2.6: The bit error probability performance of the received XORed packets before the last decoder at destination nodes under different OS selections.

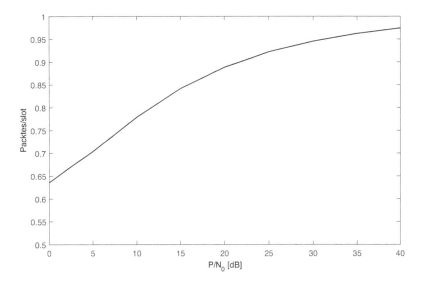

FIGURE 2.7: The average number of the received packets per slot at different destination nodes under OS.

nodes and chosen to be XORed at the relay node. This is why, in Fig. 2.6, we analyse OS selections based on the number of selected packets. In general, the higher the scheduling level k, the fewer packets are chosen to be XORed when $\mathfrak{D}_s$ $(s \in \mathfrak{S})$ is destined.

We can see from Fig. 2.6 that the derived upper bound is very tight, especially at medium and high SNR. Furthermore, regardless of which scheduling level is used in OS, the number of selected packets is always equal to or less than that of the case without OS (i.e., all the packets are selected). In other words, the scheme with OS outperforms the scheme without OS in terms of bit error probability.

In the case of OS, Fig. 2.7 depicts the simulation results for the average number of received packets per slot at different destination nodes. The vertical axis, which represents the average probability of packets received at different receiving nodes, can be interpreted as the likelihood of how many receiving nodes will be chosen as destination nodes by the OS. Furthermore, as shown in Fig. 2.7, the majority of nodes will be chosen as destination nodes at high SNR, while fewer nodes will be chosen at low SNR. With an increase in SNR, more instantaneous links will become good enough to be selected by the OS.

2.8.3 THE PERFORMANCE OF NESTED CODES

The simulation and theoretical results are shown in Fig. 2.8 when the scheduling level k is set to 1 and the RCPC code rate is set to $1/3$. In Eq. 2.42, the bit error probability considered here is $\mathcal{P}e^{s \to d}$ at a destination node. We choose an

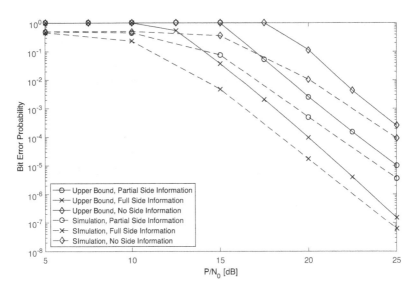

FIGURE 2.8: The performance of nested codes with different side information when OS level $k = 1$, RCPC code rate is 1/3.

arbitrary destination node and assume three different scenarios for this node: full side information, partial side information, and no side information. For example, in Fig. 2.1, full side information means that node c overhears all of its neighbors' (a, f, d) information (i_a, i_f, i_d) when they transmit to the relay, while partial side information information means node c overhears one of them, and no side information means node c only knows its own information i_c.

The node with full side information has the best performance on the bit error probability, as shown in Fig. 2.8. When it only knows its own information, on the other hand, it performs poorly. This is due to the amount of prior knowledge as well as the performance of various corresponding generator matrices. The destination nodes with the least prior knowledge, in particular, correspond to the "larger stacking" generator matrices. Furthermore, larger "stacking" generator matrices have generally poorer decoding abilities.

In addition, as shown in Fig. 2.8, the analytical upper bounds are loose at low SNR and tighten as SNR increases. This is due to the poor approximation of the upper bound over a BSC channel, as well as the imprecise estimations of the error correction ability of the stacking generator matrices.

2.9 CONCLUSIONS

We propose a multi-source multi-destination wireless relay network coded scheme in this chapter that combines the benefits of nested codes and OS. We present the proposed scheme's detailed coding process and derive upper bounds for bit error probability for schemes with and without OS. We also perform the code search based on the specified criterion. We investigate the performance of the proposed scheme for various situations in the simulation results, which show that the new network coded scheme outperforms the scheme without OS in terms of bit error probability and validates the theoretical upper bounds. In the future, we will investigate the achievable rate of our proposed system, which corresponds to RCPC codes, as well as other rate adaptive methods such as adaptive modulation and coding (AMC).

1. Y. E. Sagduyu and A. Ephremides, "Joint scheduling and wireless network coding," in *Proc. First Workshop on Network Coding, Theory, and Applications*, Riva Del Garda, Itlay, Apr. 2005.

2. H. Yomo and P. Popovski, "Oppportunistic Scheduling for Wireless Network Coding," in *Proceeding of IEEE International Conference on Communications'07*, June 2007, pp. 5610–5615.

3. L. Xiao, T. E. Fuja, J. Kliewer, and Jr D. J. Costello, "Nested codes with multiple interpretations," in *Proceeding of 2006 40th Annual conference on information sciences and systems*, Mar. 2006, pp. 851–856.

4. C. E. Shannon, "A mathematical theory of communication," *Bell System Technical Journal*, vol. 27, pp. 379–423, 623–656, July, Oct. 1948.

5. T. M. Cover and J. A. Thomas, *Elements of Information Theory*, John Wiley and Sons, Inc., 1991.

6. Q. Du and X. Zhang, "Time-sharing based rate adaptation for multicast over wireless fading channel in mobile wireless network," in *Proceeding of 2006 40th Annual conference on information sciences and systems*, Mar. 2006, pp. 1385–1390.

7. C. S. Hwang and Y. Kim, "An adaptive modulation for multicast communications of hierarchical data in wireless network," in *Proceeding of IEEE International Conference on Communications'02*, Feb. 2002, pp. 896–900.

8. A. Goldsmith and P. P. Varajya, "Capacity of Fading Channels with Channel Side Information," *IEEE Trans. Inform. Theory*, vol. 43, no. 6, pp. 1986–1992, Nov. 1997.

9. W. C. Jakes, *Microwave Mobile Communication*, IEEE Press, New Jersy, USA, IEEE reprinted edition, 1994.

10. H. A. David, *Order Statistics*, Wiley, 1980.

11. W. C. Y. Lee, "Estimate of channel capacity in Rayleigh fading enviroment," *IEEE Trans. on Veh. Technol.*, vol. 39, pp. 187–190, Aug. 1990.

12. A. Goldsmith, *Wireless Communications*, Cambridge University Press, 2005.

13. L. Xiao, T. Fuja, J. Kliewer, and D. Costello, "Nested codes with multiple interpretations," in *Proc. 40th Annual Conference on Information Sciences and Systems*, Mar. 2006, pp. 851-856.

14. P. Frenger, P. Orten, T. Ottosson, and A. Svensson, *Multi-rate convolutional codes*, Chalmers University of Technology Press, 1998.

15. Z. Lin and B. Vucetic, "Power and rate adaptation for wireless network coding with opportunistic scheduling," in *The IEEE International Symposium on Information Theory*, Toronto, Jul. 6–11, 2008.

16. M. Simon and M. Alouini, *Digital Communication over Fading Channels – A Unified Approach to Performance Analysis*, First Edition., Wiley, 2000.

17. J. Hagenauer, "Rate-compatible punctured convolutional codes (RCPC codes) and their applications," *IEEE Trans. Commun.*, vol. COM-36, pp. 389–400, Apr. 1988.

18. F. Zhao and Z. Bai, "Simplified BER analysis of rate-compatible punctured convolutional coded cooperative system over slow Rayleigh fading channel," in *Proc. ISCIT'09*, Icheon, pp. 943–946, Sep. 2009.

19. J. Proakis, *Digital Communications (Fourth Edition)*, McGraw-Hill, 2006.

20. T. Moon, *Error Correction Coding – Mathematical Methods and Algorithms*, Wiley-Interscience, 2005.

21. J. Chang, D. Hwang and M. Lin, "Some extended results on the search for good convolutional codes," *IEEE Trans. Inform. Theory*, vol. 43, no. 5, pp. 1672–1697, Sep. 1997.

3 Distributed Network Coded Modulation Schemes for Multiple Access Relay Channels

3.1 INTRODUCTION

Distributed coding, as a special channel coding strategy developed for cooperative communication networks [1, 2], attracted large attentions recently. The distributed codes construction concept has been applied on conventional channel coding to form such as distributed turbo codes [3], distributed space-time codes [4], and distributed low-density parity-check (LDPC) codes [5]. These published results show that the proposed schemes can improve the transmission reliability over point-to-point wireless communication channels.

The distributed coding schemes discussed above are developed for small-scale unicast relay networks, in which the messages are sent from a single source to a single destination through single/multi-hop relays. In this work, we consider a scenario that a source Mobile Terminal (MT) cooperating with other MTs for uplink transmission to the Base Station (BS) via a Relay Node (RN). A classical way to pass such kind of information is through routing, where the relay nodes simply store and forward the received packets to the destination.

Recently, a network coding (NC) [6] approach is proposed to replace routing. In NC, the relay nodes are allowed to encode the packets received from multiple source nodes. The combined information is subsequently sent to the destination. It has been shown in [6] that compared with traditional routing, NC can enhance the network capacity and throughput.

Given the research in distributed network-channel codes (DNCC) design started recently, many open questions in the design and implementation of distributed codes still have not been addressed. In this work, we develop the design of the distributed codes for cooperative uplink transmissions in cellular systems and analyze the code performance based on the framework of uplink cellular systems.

The main contributions of this work are the propose of a distributed physical layer network coded systems, the derivation of the analytical upper bound on the error probability and the propose of an iterative decoding approach for the interleaved distributed coded system.

DOI: 10.1201/9781003203803-3

3.2 SYSTEM MODEL

We consider a wireless cellular uplink transmission system, in which a MT sends data packets to a BS via a RN. The source MT can collaborate with other MTs for cooperative transmission. We assume that there are direct links between the source MT and the cooperative MTs, but no direct links between the BS and the MTs. The discussion of the assumption of the direct links between the MTs is given in [7]. The transmission can be separated into two time slots, in the first slot, the source MT first encodes the information packets then broadcasts to both the RN and cooperative MTs. Both the RN and cooperative MTs begin to decode. If the RN can correctly decode the codewords from the source MT, then the codewords will be re-encoded, modulated and broadcasted to the BS. In this case, no cooperative phase is needed. Otherwise, the RN will ask the cooperative MTs for transmission of the same message from the source MT. The cooperative MTs will use different encoders from the source MT. At the RN, the received waveform from both the source and cooperative MTs are symbol wise alternatively concatenated and detected based on the joint encoders for both the source and cooperative MTs. Although multiple cooperative MTs can be employed in the system, in this work, we only consider the case of using one single cooperative MT.

The system model is illustrated in Fig. 3.1, in which a source MT cooperating with a cooperative MT transmits information packets to a BS via a RN. In the cooperative phase, the source and cooperative MTs use the encoder C_1 with a code rate of $1/n_1$ and C_2 with a code rate of $1/n_2$, respectively. At the RN, the received waveforms from both the source and cooperative MTs are concatenated symbol wise alternatively, which is similar to the operation of physical layer network coding [8]. Instead of mixing the two received waveforms as in [8], in this work, the received waveforms are first separated based on the symbol interval, then the separated waveforms for every n_1 symbol intervals from the source MT and every n_2 symbol intervals from the cooperative MTs are alternatively concatenated to form a network coded waveform. Since the waveforms are concatenated in the air, it is also termed as physical layer network coded signal. The network coded waveform is first demodulated and then decoded with a decoder for the joint encoder of C_1 and C_2 from the source and cooperative MTs, respectively. At the RN, the decoder decodes the physical layer network coded signals based on the maximum likelihood sequence detection (MLSD) algorithm, the output of the decoder is the estimated codewords of the joint encoder of C_1 and C_2. The output codewords are then encoded, modulated and transmitted. Since the encoder encodes the signals from both the source and cooperative MTs, it in fact performs a network coding operation.

3.3 DISTRIBUTED NETWORK CODED MODULATION SCHEMES BASED ON PUNCTURED CONVOLUTIONAL CODES

The source MT first encodes the information packets into codewords then broadcasts to the RN and the cooperative MT. In this work, we consider a class of Rate Compatible Convolutional (RCC) encoders [9–11] to encode the source information

Distributed Network Coded Modulation Schemes

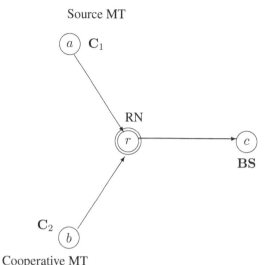

FIGURE 3.1: An uplink transmission system with one source MT, a cooperative MT, a Relay node and a BS.

at both of the source and cooperative MTs. RCC codes are a family of codes in which all coded bits of high rate codes are obtained by the lower rate codes. That is, lower rate codes use all the symbols of the high rate codes plus some extra redundancy symbols. This allows transmission of incremental redundancy in ARQ/FEC schemes and continuous rate variation, changing from low to high error protection within a data frame. A very powerful and flexible RCC coding scheme is obtained by combing puncturing and repetition; puncturing for high rate codes and repetition for low rate codes [9].

After receiving the codewords, both the RN and the cooperative MT begin to decode. The cooperative MT then decode the message from the source MT, and re-encode the decoded message using a different RCC encoder from the one used by source MT. The codeword from the cooperative MT is then sent to the RN at the second time slot. Suppose that each encoder at the source MT and cooperative MTs has the same rate of k/n. Then for a system with a source MT and a cooperative MT, the joint encoder for all MTs has a rate of $k/2n$. We assume that the encoders for all MTs have the same constraint length [12].

For a rate $1/n$ convolutional encoder, the generator polynomial matrix is [13]

$$G_N(D) = \begin{bmatrix} G^0(D) \\ \vdots \\ G^{n-1}(D) \end{bmatrix}, \text{ where } G^i(D) = g_0^i + g_1^i D + \cdots + g_m^i D^m \text{ for } i \in$$

$\{0, \ldots, n-1\}$. The coefficients g_j^i for $0 \leq j \leq m$ belong to the set $\{0, 1\}$. A high rate Punctured Convolutional Code (PCC) can be obtained by puncturing a parent $1/n$ binary convolutional code. The operation of puncturing some coded symbols is implemented by using an $(n \times p)$ puncturing matrix, P_{mat}, where p is the puncturing period. Let s be the total number of transmitted bits during a puncturing period p, the coding rate of a PCC is $r = p/s$. For two punctured convolutional codes obtained from the same parent code, we call them rate compatible if the higher rate code is obtain by puncturing the lower rate code according to the rate compatibility criterion. This criterion requires that all the lower rate code make use of all code bits of the high rate code, plus one or more additional bits. In other words, the puncturing matrix of the high rate code is embedded into the puncturing matrix of the lower rate code.

The notion of repetition convolutional code was introduced in [14]. The construction method for repetition codes is similar to the one for punctured codes. The difference is that instead of puncturing the coded bits for punctured codes, the repetition codes are generated by duplicating some code symbols of a rate $1/n$ parent convolutional code. The operation of duplicating the coded bits is done by an $(n \times p)$ repetition matrix (where p is repetition period). But now the elements of this matrix are greater than or equal to one, indicating the number of duplications of that corresponding coded bit. The starting rate for Rate Compatible Repetition Convolutional (RCRC) codes is determined by n and p. Clearly, the repetition code obtained from a noncatastrophic code cannot be catastrophic. However, one can choose from among all possible combinations of encoded bits to be duplicated which yields the best repetition code. Two repetition codes obtained from the same parent code are considered to be rate compatible if the lower rate code uses all the coded bits of the higher rate code, and plus one or more duplicated bits. This means that all the elements in the repetition matrix of the lower rate code must be greater than or equal to the corresponding elements in the repetition matrix of the higher rate code. For example, for a rate $2/5$ repetition convolutional code of memory $m = 6$ and repetition period $p = 2$, its optimum repetition matrix is given by [11]

$$Q_{2/5} = \begin{bmatrix} 2 & 1 \\ 1 & 1 \end{bmatrix}$$

and the matrix for a rate compatible repetition convolutional code of rate 2/6 can be any of the matrices

$$Q_{2/6} = \begin{bmatrix} 2 & 2 \\ 1 & 1 \end{bmatrix}, \begin{bmatrix} 2 & 1 \\ 2 & 1 \end{bmatrix}, \begin{bmatrix} 2 & 1 \\ 1 & 2 \end{bmatrix}, \text{or} \begin{bmatrix} 3 & 1 \\ 1 & 1 \end{bmatrix}.$$

The encoded symbols from the MTs are BPSK modulated and then transmitted to the RN with different time slots. At the RN, the received signals from the source MT and the cooperative MT are jointly decoded with a decoder corresponding to the joint encoder of both the source and cooperative MTs. The output of the decoder is the estimated codewords of the joint encoder. It is the encoded codewords rather than the information block of the source MT. In this work, we employ the hybrid type II ARQ scheme [9] in the links between MTs and the RN to ensure that the codewords output from the joint decoder at the RN can be estimated without any error. This can be done by appending a Cyclic Redundancy Check (CRC) [15] to each transmitted data packet of the source MT. If the decoded information date packet is failed with the

CRC, the RN will request the MTs to transmit the punctured coded bits or repetition coded bits at the MTs according to the rate compatible criterion [11]. This ARQ scheme ensures that the codewords of the joint distributed encoders for both MTs can be successfully decoded at the RN. The output codewords are then fed into a Network Encoder (NE) and modulated with a memoryless modulator (MM). By properly choosing the NE and the MM, we can form a class of digital coded modulation scheme. To simplify the study, here we borrow the idea of coded digital phase modulation scheme, such as Continuous Phase Frequency Shift Keying (CPFSK) [12, 18], in which the NE is a Recursive Systematic Convolutional (RSC) encoder. It was shown in [16] that a CPFSK scheme can be decomposed into a concatenation of a RSC and a MM. CPFSK has the advantage of creating a differentially encoded waveform stream and is further attractive due to its good spectral properties.

The generated waveform from the MSK modulator is transmitted over the channel. The channel is assumed to be the AWGN channel. At the RN, the NE takes one output symbol from the joint MTs decoder as an input and generates one vector which is used by the MM to generate one channel transmission waveform.

3.3.1 DECODING WITH NETWORK CODED MODULATION AT THE DESTINATION NODE

The received passband signal at the destination node can be expressed as, $r(t, \boldsymbol{\mu}) = s(t, \boldsymbol{\mu}) + n(t)$, where $n(t)$ is a zero mean Gaussian random process. $s(t, \boldsymbol{\mu})$ is the transmitted signal from the relay with binary information symbols $\mu \in \{0, 1\}$, which can be written as [16]

$$s(t, \boldsymbol{\mu}) = \mathrm{Re}\left\{s_b(t, \boldsymbol{\mu})e^{j(2\pi f_1 t + \phi_0)}\right\} \quad (3.1)$$

where $s_b(t, \boldsymbol{\mu}) = \sqrt{\frac{2E_s}{T}}e^{j\overline{\psi}(t,\boldsymbol{\mu})}$ is the complex baseband equivalent signal [17]. $f_1 = f_c - h/2T$ is a shift of the carrier frequency f_c, ϕ_0 is the initial phase of the carrier and $\boldsymbol{\mu}$ is the data symbol sequence. E_s and T are the symbol energy and the symbol interval duration, respectively.

The *tilted information carrying phase*[1] during symbol interval n ($t = \tau + nT$) is given by (3.2),

$$\overline{\psi}(\tau + nT, \boldsymbol{\mu}) = R_{2\pi}\left\{2\pi h R_P\left\{\sum_{i=0}^{n-L} \mu_i\right\} + 4\pi h \sum_{i=0}^{L-1} \mu_{n-i} q(\tau + iT) + W(\tau)\right\},$$
$$0 \leq \tau < T \quad (3.2)$$

---
[1] In [16] representation of CPM.

where $R_x\{\cdot\}$ is the modulo x operator and $W(\tau)$ is given by (3.3), which

$$W(\tau) = \pi h(M-1)\frac{\tau}{T} - 2\pi h(M-1)\sum_{i=0}^{L-1} q(\tau+iT) + \pi h(M-1)(L-1),$$
$$0 \leq \tau < T \tag{3.3}$$

represents the data-independent terms. The phase response $q(\cdot)$ is found from the frequency pulse $g(t)$ according to $q(t) = \int_{-\infty}^{t} g(\tau)d\tau$. L is the length of the frequency response. For CPFSK, $L = 1$ and the frequency pulse used in this work is the Rectangular (REC) pulse defined as:

LREC $\quad g(t) = \frac{1}{2T}, \quad 0 \leq t \leq T$

The equivalent complex baseband continuous-time signal can be written as

$$r_b(t, \boldsymbol{\mu}) = s_b(t, \boldsymbol{\mu}) + n_b(t), \tag{3.4}$$

where $n_b(t)$ is a complex baseband representation of the additive white Gaussian random process having zero mean double-sided power spectral density $2N_0$ [17, 18].

In each symbol interval, a set of possible CPM sequences consists of $P \cdot M^L$ various complex valued signals. Thus, a bank of $P \cdot M^L$ complex valued filters, matched to those signals and sampled once every symbol interval produces a sufficient statistics, since they form a basis for signal space [17]. A component of the vector $\mathbf{r}_n$ representing the sampled outputs at symbol interval n, can be calculated by

$$r_{i,n} = \int_{(n-1)T}^{nT} r(t, \boldsymbol{\mu})e^{j\overline{\psi}(t,\tilde{\boldsymbol{\mu}}_i)}dt, \tag{3.5}$$

where $i \in \{1, 2, \ldots, P \cdot M^L\}$ and $\tilde{\boldsymbol{\mu}}_i$ is the ith hypothesis sequence offered by the receiver. All the $P \cdot M^L$ various complex signals must be generated by using various $\tilde{\boldsymbol{\mu}}$ s and the index i on the right hand side of (3.5) is intended to reflect this.

Sufficient statistics can be produced in many other ways, corresponding to choosing other basis functions for the signal space. Changing from one to another is done by a linear transformation. Given the above choice, signal space basis may not be orthogonal. There can be linear deterministic dependencies among the components of $\mathbf{r}_n$. Hence $\mathbf{r}_n$ has a non-diagonal covariance matrix $\boldsymbol{\Lambda}$.

We denote by $\boldsymbol{\chi}_n$ the expectation of $\mathbf{r}_n$. Because of the Gaussian channel, $\mathbf{r}_n$ is a Gaussian random vector. The joint Probability Density Function (PDF), conditioned on the CPE output vector $\boldsymbol{\nu}_n$, is [18]

$$p(\mathbf{r}_n|\boldsymbol{\nu}_n) \propto \exp\left\{-(\mathbf{r}_n - \boldsymbol{\chi}_n)^H \boldsymbol{\Lambda}^{-1}(\mathbf{r}_n - \boldsymbol{\chi}_n)\right\} \tag{3.6}$$

where $(\cdot)^H$ denotes Hermitian transpose. A more detailed description of CPM can be found in [18, 21].

Based on (3.6), we can use maximum a posterior (MAP) algorithm to decode the received CPFSK signal as described in [18]. Please note that since the generators at the source nodes and the CPFSK modulator formed a super-trellis at the destination nodes, we can employ the MAP algorithm over the super-trellis to decode the transmitted signal for each source node.

3.3.2 ANALYTICAL BOUNDS ON THE BIT ERROR PROBABILITY FOR THE MULTIPLE ACCESS RELAY CHANNELS

In this work, we will develop an analytical upper bound on the distributed coded system under MLSD. The distributed convolutional encoders for MTs and the NE at the RN constitute a super-trellis encoder. At the BS, we can develop a ML decoder for the super-trellis encoder. The state σ_j of the super-trellis encoder at discrete time j is defined as $(\sigma_j^{cc}, \sigma_j^{ne})$, where σ_j^{cc} and σ_j^{ne} denote the state of the Joint Distributed punctured Convolutional Encoder (JDCE) and the state of RSC at discrete time j, respectively. For a JDCE having m memory elements, and a CPFSK scheme, the total number of states is $2^{(m+1)}$. The state transition $\sigma_j \to \sigma_{j+1}$ is determined by the input of the source MT. Associated with this transition is also the input symbol $\mu \in \{0, 1\}$ of the NE, i.e., the RSC encoder, and the mean vector which is obtained by letting the transmitted waveform pass through a bank of complex filters which are matched to the transmitted signals [18].

In this work, we assume both the source and the cooperative MT have the same transmit power. Let E_b be the information bit energy and $N_0/2$ be the double sided power spectral density of the additive white Gaussian noise. The bit error probability for a memoryless information source sequence of the distributed coded modulation system will follow the following theorem.

Theorem 1. *Under the MLSD and the assumption that the source block is infinitely long, the bit error probability for a distributed rate compatible convolutional encoded CPFSK system with a discrete memoryless uniform digital source sequence, can be upper bounded by*

$$P_b < Q\left(\sqrt{d_{min}^2 \frac{E_b r}{N_0}}\right) \exp\left(d_{min}^2 \frac{E_b r}{2N_0}\right) \cdot \frac{\partial F(\eta, \epsilon, \zeta)}{\partial \epsilon} \bigg|_{\eta=1/2, \epsilon=1, \zeta=e^{(-E_b r/2N_0)}}, \quad (3.7)$$

where d_{min}^2 is the minimum NSED and η, ϵ, ζ are dummy variables and r is the code rate of the joint trellis encoder of the source and cooperative MTs, $r = p/s$ as described in Section II. The average transfer function is

$$F(\eta, \epsilon, \zeta) = M^{-m} \sum_{\kappa=1}^{M^m} F(j, \kappa, \eta, \epsilon, \zeta) = M^{-m} \frac{1}{p} \sum_{j=0}^{p-1} \sum_{\kappa=1}^{M^m} \sum_{\iota} \sum_{\tau} \sum_{d} W_{j, s_\kappa, \iota, \tau, d} \eta^\iota \epsilon^\tau \zeta^{d^2} \quad (3.8)$$

where $W_{j, s_\kappa, \iota, \tau, d}$ is the number of error events that start at time j from state s_κ, and have NSED d^2, length ι and total number of symbol errors caused by the error event given by τ. The Q function is defined as $Q(x) = (\sqrt{2\pi})^{-1} \int_x^\infty e^{-z^2/2} dz$.
□

Proof. The decoding process is performed at the receiver of the BS. Denote by ς_j the total bit error caused by all error events starting at a discrete time j. Let us take an arbitrary state, say state s, $s \in S$, where S is the state space. Let $\Xi^e_{j,s,l,\tau,d}$ be the error event starting at time j with initial state s, length l, and Normalized Squared Euclidean Distance (NSED) d^2 [21]. τ represents the symbol error caused by the error event $\Xi^e_{j,s,l,\tau,d}$. $\Xi^e_{j,s,l,\tau,d}$ is completely described by the start state s and the pair sequences $(\boldsymbol{y}_{s,l}, \hat{\boldsymbol{y}}_{s,l})$, resulting in $\Xi^e_{j,s,l,\tau,d}$. Here $\boldsymbol{y}_{s,l}$ and $\hat{\boldsymbol{y}}_{s,l}$ are the reconstructed data sequences.

For a distributed punctured or repetition convolutional coded CPFSK system, the NSED d^2 associated with an error event $\Xi^e = \boldsymbol{\mu} - \boldsymbol{\mu}'$ can be calculated as

$$d^2 = r \cdot \left(1 - \frac{1}{T}\sum_{i=0}^{l-1}\int_{iT}^{(i+1)T} \cos\phi(t,\boldsymbol{\gamma})dt\right). \tag{3.9}$$

Here T is the symbol interval duration and $\phi(t,\boldsymbol{\gamma}) = [\pi\omega_i + 2\pi\gamma_i q(t-iT)]$, where $\omega_j = R_P\{(\sigma^{rsc}_j - \hat{\sigma}^{rsc}_j)\} = R_P\left\{\sum_{n=0}^{j-1}\gamma_n\right\}$ is the difference phase state, and $\gamma_i \in \{-2,0,2\}$. $R_x\{\cdot\}$ is the modulo x operator and $q(t) = \frac{t}{2T}$ is the phase response [21].

The expected value of the bit error rate, caused by error events starting at time j, is given by

$$\begin{aligned}E[\varsigma_j] &= \sum_s\sum_l\sum_\tau\sum_d W_{j,s,l,\tau,d}\cdot\tau\cdot Pr(\Xi^e_{j,s,l,\tau,d})\\ &= \sum_s\sum_l\sum_\tau\sum_d W_{j,s,l,\tau,d}\cdot\tau\cdot Pr(\hat{\boldsymbol{Y}}_j = \hat{\boldsymbol{y}}_{s,l}|\boldsymbol{Y}_j = \boldsymbol{y}_{s,l})\cdot Pr(\boldsymbol{Y}_j = \boldsymbol{y}_{s,l}),\end{aligned} \tag{3.10}$$

where $\boldsymbol{Y}_j$ and $\hat{\boldsymbol{Y}}_j$ are two random vectors, whose outcome space are all possible reconstructed signal sequences starting at time j. $W_{j,s,l,\tau,d}$ is the number of error events that start with state s, and have NSED d^2, length l and total error symbols τ. The expectation of (3.10) is over all error events starting at time j.

Suppose that the encoding rate of the parent convolutional encoder is k/n. Then there are 2^k branches entering and leaving each state. For a discrete memoryless uniform digital source, all sequences that start with a state s and have length l, are equally probable. Thus, we can write

$$Pr(\boldsymbol{Y}_j = \boldsymbol{y}_{s,l}) = Pr(\sigma_j = s)\cdot 2^{-k\cdot l}. \tag{3.11}$$

The conditional probability $Pr(\hat{\boldsymbol{Y}}_j = \hat{\boldsymbol{y}}_{s,l}|\boldsymbol{Y}_j = \boldsymbol{y}_{s,l})$ can be upper bounded by $Q(\sqrt{\frac{d^2 E_b r}{N_0}})$, where the Q function is defined as $Q(x) = (\sqrt{2\pi})^{-1}\int_x^\infty e^{-\tau^2/2}d\tau$. This can be proved as follow: Let Φ be the set of all sequences $\boldsymbol{Y}_{s,l}$ starting at time j with state s having length l. Denote by $\|\Phi\|$ the cardinality of the set Φ. When $\|\Phi\| = 2$, there are only two sequences in the set Φ, $\boldsymbol{y}_{s,l}$ and $\hat{\boldsymbol{y}}_{s,l}$. With

MLSD, the decision region of $\hat{y}_{s,l}$ is half of the signal space and the error probability $Pr(\hat{Y}_j = \hat{y}_{s,l} | Y_j = y_{s,l})$ is exactly given by $Q(\sqrt{2d^2 E_b r / 2N_0})$, where d^2 is the NSED between the transmitted signals $s(t, y_{s,l})$ and $s(t, \hat{y}_{s,l})$. Therefore, $Pr(\hat{Y}_j = \hat{y}_{s,l} | Y_j = y_{s,l}) = Q(\sqrt{d^2 E_s r / (N_0)})$. When $\|\Phi\| > 2$, the decision region of $\hat{y}_{s,l}$ is less than half of the decision space. Therefore, $Pr(\hat{Y}_j = \hat{y}_{s,l} | Y_j = y_{s,l}) < Q(\sqrt{d^2 \frac{E_b r}{N_0}})$.

With the above result and (3.11), by considering all error events and using the union bound technique [13], $E[\varsigma_j]$ can be upper bounded by

$$E[\varsigma_j] \leq \sum_s Pr(\sigma_j = s) \sum_l \sum_\tau \sum_d W_{j,s,l,\tau,d} \cdot \tau \cdot 2^{-k \cdot l} Q(\sqrt{d^2 \frac{E_b r}{N_0}}). \quad (3.12)$$

To further reduce the complexity of the computation of (3.12), one need to note that for a coded CPFSK system, some of the states are equivalent. Here "equivalent" means that if the error events Ξ_σ^e and Ξ_ϱ^e, starting at time j, generated by the same pair sequences $(y_l, \hat{y}_l)$ starting at states σ and ϱ, respectively, are identical. Here "identical" means that the error events have the same length, the same NSED, and generate the same number of symbol errors. In the following, we mathematically denote two equivalent states σ and ϱ by $\sigma \equiv \varrho$.

Based on the above statement, all the states of the punctured or repetition convolutional coded CPFSK at time j, having the same state of the encoder, σ_j^{cc}, are equivalent. Among the 2^{m+1} states of the coded CPFSK, there are only 2^m distinct states which are not equivalent to each other. Consequently, (3.12) can be written as

$$E[\varsigma_j] \leq \sum_{\kappa=1}^{2^m} Pr(\sigma_j \equiv s_\kappa) \sum_l \sum_\tau \sum_d W_{j,s_\kappa,l,\tau,d} \cdot \tau \cdot 2^{-k \cdot l} Q(\sqrt{d^2 \frac{E_b r}{N_0}}). \quad (3.13)$$

For an i.i.d. discrete memoryless source, the encoder can start at any one of the 2^m trellis states. Further, since the trellis is periodically time variant, $E[\varsigma_j]$ is periodically dependent on the time j. The probability that a state σ_j of the coded CPFSK at time j is equivalent to one of the 2^m distinct states is

$$Pr(\sigma_j \equiv s_\kappa) = 2^{-m}, \quad \forall j. \quad (3.14)$$

Under the assumption that the source sequence is infinitely long, by using the inequality $Q(\sqrt{x+y}) \leq Q(\sqrt{x}) e^{-y/2}$ for $x, y > 0$ [21], and letting $d^2 = d_{min}^2 + (d^2 - d_{min}^2)$, the symbol error probability can be upper bounded by

$$P_b \leq Q\left(\sqrt{d_{min}^2 \frac{E_b r}{N_0}}\right) \exp\left(d_{min}^2 \frac{E_b r}{2N_0}\right) 2^{-m} \frac{1}{p} \sum_{j=0}^{p-1} \sum_{\kappa=1}^{2^m}$$
$$\sum_l \sum_\tau \sum_d W_{j,s_\kappa,l,\tau,d} \cdot \tau \cdot 2^{-k \cdot l} \cdot \exp(-d^2 \frac{E_b r}{2N_0}), \quad (3.15)$$

where p is the puncturing or repetition period as described in Section 3.3, and $j = 0$ corresponding the beginning of a puncturing or repetition period.

The right hand side of the inequality (3.15) can be computed by means of the generating function. Let us denote by $F(j, \kappa, \eta, \epsilon, \zeta)$ the generating function for one of the distinct equivalent states s_κ, $\kappa = 1, 2, \ldots, 2^m$. $F(j, \kappa, \eta, \epsilon, \zeta)$ is then given by

$$F(j, \kappa, \eta, \epsilon, \zeta) = \sum_l \sum_\tau \sum_d W_{j, s_\kappa, l, \tau, d} \eta^l \epsilon^\tau \zeta^{d^2} \qquad (3.16)$$

where η, ϵ, ζ are "dummy" variables [13]. Using the above equation, Eq. (3.15) can be written as

$$P_b \leq Q\left(\sqrt{d_{min}^2 \frac{E_b r}{N_0}}\right) \exp\left(d_{min}^2 \frac{E_b r}{2N_0}\right) \cdot$$
$$\frac{\partial F(\eta, \epsilon, \zeta)}{\partial \epsilon}\bigg|_{\eta = 2^{-k}, \epsilon = 1, \zeta = e^{(-E_b r / 2N_0)}}, \qquad (3.17)$$

where the average generating function $F(\eta, \epsilon, \zeta)$ is given by

$$F(\eta, \epsilon, \zeta) = 2^{-m} \frac{1}{p} \sum_{j=0}^{p-1} \sum_{\kappa=1}^{2^m} F(j, \kappa, \eta, \epsilon, \zeta). \qquad (3.18)$$

□

The transfer function $F(j, \kappa, \eta, \epsilon, \zeta)$ can be obtained by using a product state diagram [19, 20]. A product state at time j is defined as $(\sigma_j, \hat{\sigma}_j)$, where σ_j is a state of the super-trellis encoder of the distributed system and $\hat{\sigma}_j$ represents a state of the decoder. The transition $(\sigma_j, \hat{\sigma}_j) \to (\sigma_{j+1}, \hat{\sigma}_{j+1})$ is labeled with

$$\sum_{\Delta \tau} \sum_{\Delta d^2} b(\Delta \tau, \Delta d^2) \eta \epsilon^{\Delta \tau} \zeta^{\Delta d^2}, \qquad (3.19)$$

where $\Delta \tau$ and Δd^2 are the number of the symbol errors and NSED, respectively. $b(\Delta \tau, \Delta d^2)$ denotes the number of paths having NSED Δd^2 and symbol errors $\Delta \tau$ for this state transition.

For the distributed RCC coded CPFSK system, the bit error rate only depends on the output of the RCC encoder. In other words, it is independent of the pair state of the RSC within the CPFSK modulator $(\sigma_j^{ne}, \hat{\sigma}_j^{ne})$. Furthermore, the NSED d^2 only depends on the difference of the CPFSK states $(\sigma_j^{ne} - \hat{\sigma}_j^{ne})$ [18]. Therefore, the product state can be reduced. The reduced product state can be written as $(\sigma_j^{cc}, \hat{\sigma}_j^{cc}, \omega_j)$, where $\omega_j = R_P\left\{(\sigma_j^{ne} - \hat{\sigma}_j^{ne})\right\} = R_P\left\{\sum_{n=0}^{j-L} \gamma_n\right\}$ is the difference phase state, see (3.9). The total number of product states for the distributed RCC coded CPFSK systems is 2^{2m+1}.

The product states can be divided into initial states, transfer states, and end states. A product state is an initial state if an error event can start from it. A product state is an end state if an error event can end in it. The conditions for initial states and end states are $\sigma_j^{cc} = \hat{\sigma}_j^{cc}$ and $\omega_j = 0$. Other states are referred to as transfer states.

Let $A_{\kappa,j}$ represent the state transitions from an initial state s_κ to transfer states in one step at time j. Let us denote by B_j the transitions from transfer states to end states, and by C_j the transitions from transfer states to transfer states in one step at time j. Let $D_{\kappa,j}$ represent the transitions from an initial state s_κ to end states in one step. The transfer function can be calculated by [21]

$$F(j,\kappa,\eta,\epsilon,\zeta) = \mathbf{1} \cdot \left(B_j (I - C_j)^{-1} A_{\kappa,j} + D_{\kappa,j} \right) \qquad (3.20)$$

where $\mathbf{1}$ is an all one vector and I represents the identity matrix.

Equation (7) can be further expressed as

$$P_b \leq \sum_{d^2} W_d \cdot Q(\sqrt{d^2 \frac{E_b r}{N_0}}), \qquad (3.21)$$

where $W_d = \frac{2^{-m}}{p} \sum_{j=1}^{p} \sum_{\kappa=1}^{2^m} \sum_l \sum_\tau W_{j,s_\kappa,l,\tau,d} \cdot \tau \cdot 2^{-l}$.

It can be seen from (3.21) that the minimum NSED d_{min}^2 and $W_{d_{min}}$ (the number of error events with d_{min}^2) dominate the asymptotical symbol error rate of the system.

Largest d_{min}^2 is used as the design criterion for the selection of the best puncture ore repetition matrices of the investigated distributed network coded systems.

3.4 INTERLEAVED DISTRIBUTED NETWORK CODED SYSTEMS

In the above section, the data packets decoded from the joint RCC decoder at the RN will be directly network encoded and modulated with CPFSK. In this section, we propose an interleaved network encoded scheme, in which the decoded data packet at the RN will be first interleaved then network encoded with a NE and modulated with a memoryless modulator, here we still consider a CPFSK modulator. At the BS, when receiving the interleaved network coded signals, the BS will perform iterative decoding algorithm to decode the message from the source MT. A block diagram of the system is shown in Fig. 3.2. The subscript k denotes discrete time. π denotes a random interleaver and π^{-1} denotes the deinterleaver.

The iterative algorithm consists of one Soft-In Soft-Out (SISO) algorithm [22] for the RSC encoder and one for the joint distributed RCC encoder of the MTs. The joint distributed RCC encoder is either a punctured or a repetition encoder, it is the one who gives the correctly decoded codeword at the RN. For the resulted RCC encoder whether it is a punctured or a repetition convolutional encoder, the trellis of the encoder is periodically time varying. Let $\tilde{\mathbf{Y}}_k = (\tilde{Y}_k^1, \tilde{Y}_k^2, \ldots, \tilde{Y}_k^{n^0})$ be the output symbols of the joint encoder at time k, where $n^0 \leq n$. Denote by $\tilde{B}_k$ the corresponding input symbol. Let $\tilde{b}_k$ and $\tilde{y}_k^j$ be a realization of $\tilde{B}_k$ and $\tilde{Y}_k^j$, respectively, and $\tilde{b}_k, \tilde{y}_k^j \in \{0,1\}$, $j \in \{1,\ldots,n^0\}$. For a trellis section at discrete time k, we define the log ratio of the APP values for the jth symbol of $\tilde{\mathbf{Y}}_k$ as

$$\Lambda_{k,1}^{j,O} = \log \frac{\sum_{e:\tilde{Y}_k^j(e)=1} \alpha_{k-1}(s') \cdot \Gamma_k(s',s) \cdot \beta_k(s)}{\sum_{e:\tilde{Y}_k^j(e)=0} \alpha_{k-1}(s') \cdot \Gamma_k(s',s) \cdot \beta_k(s)}. \qquad (3.22)$$

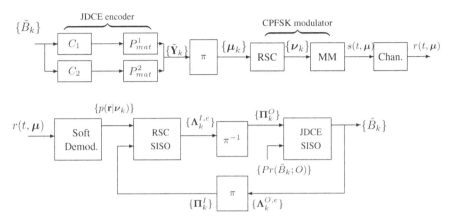

FIGURE 3.2: Iterative decoding for interleaved distributed network coded systems.

Here, $\tilde{Y}_k^j(e)$ is the jth symbol of $\tilde{\mathbf{Y}}_k$ of edge e which is associated with the state transition from s' to s at time k and letter O denotes the outer decoder. $\alpha_k(s)$ and $\beta_k(s)$ can be recursively computed from the branch metric $\Gamma_k(s', s)$ as in [23] $\alpha_k(s) = \sum_{s'} \alpha_{k-1}(s')\Gamma_k(s', s)$ and $\beta_k(s') = \sum_s \beta_{k+1}(s)\Gamma_{k+1}(s', s)$.

The branch metric $\Gamma_k(s', s)$ for trellis section at time k can be computed by $\Gamma_k(s', s) = Pr(\tilde{B}_k = \tilde{b}_k) \prod_{j=1}^{n^0} Pr(\tilde{Y}_k^j = \tilde{y}_k^j)$. For trellis sections without puncturing or repetition, $n^0 = n$. Let $\Pi_{k,i}^{j,O}$ be the log ratio of the *a priori* values for jth symbol of $\tilde{\mathbf{Y}}_k$, then $\Pi_{k,1}^{j,O} = \log \frac{Pr(\tilde{Y}_k^j=1)}{Pr(\tilde{Y}_k^j=0)}$, Eq. (3.22) can be further expressed as $\Lambda_{k,1}^{j,O} = \Lambda_{k,1}^{j,O,e} + \Pi_{k,1}^{j,O}$ where $\Lambda_{k,1}^{j,O,e}$ is the extrinsic information for the jth symbol of $\tilde{\mathbf{Y}}_k$. It is obtained from the other symbols rather than the symbol j itself, that is why we call it extrinsic information.

With the iterative decoding algorithm [22], the log ratio of the extrinsic APP values $\Lambda_k^{j,O,e} = \Lambda_{k,1}^{j,O,e}$ of the outer joint RCC decoder is passed to the inner RSC SISO module as the *a priori* information $\Pi_k^{j,I} = \Pi_{k,1}^{j,I}$ for the next iteration, the letter I denotes the inner decoder. Similarly, the extrinsic APP values $\Lambda_k^{j,I,e}$ from the inner decoder becomes the *a priori* information $\Pi_k^{j,O}$ for the outer RCC decoder in the next iteration.

The convergence behavior of the proposed systems can be analyzed with the EXIT chart [24]. The EXIT chart technique is based on the assumption that with a large random interleaver, the Log-Likelihood Ratio (LLR) of the extrinsic information passed from one SISO to the other can be modeled as an independent Gaussian random variable [25]. Through independent simulations of the two SISOs, the input/output mutual information with parameter E_s/N_0 can be estimated [24, 26].

3.5 SIMULATION RESULTS FOR DISTRIBUTED NETWORK CODED SYSTEMS

3.5.1 SIMULATION RESULTS FOR DISTRIBUTED NETWORK CODED SYSTEM WITHOUT INTERLEAVER

Computer simulations have been performed for the distributed physical layer network coded and digital phase modulated systems over AWGN channels. Several different systems with different codes are simulated. In the first system, the encoder of the source MT is obtained by puncturing a rate 1 parent non-systematic convolutional code with the generator polynomial matrix $G(D) = [1 + D + D^2]$. The encoder of the cooperative MT is obtained by puncturing a rate 1 parent non-systematic convolutional code with the generator polynomial matrix $G(D) = [1 + D^2]$. Therefore, the joint encoder of the source and the cooperative MTs is obtained by puncturing a rate $1/2$ parent non-systematic convolutional code with the generator polynomial matrix $G(D) = [1 + D + D^2; 1 + D^2]$. Table 3.1 shows the puncture matrices for the investigated network coded MSK (i.e., CPFSK with 1REC and $h = 1/2$) system [12]. The puncturing matrices in Table 3.1 are given by P_{mat}, which are obtained by an exhaustive search from all of the possible puncturing patterns for the one which gives the largest minimum NSED. The puncturing matrix is given in octal form, for example, $(3,5)_o$ represents the puncture matrix $[011; 101]$ where 0 corresponds the position that the symbol is punctured. Also shown in Table 3.1 is the observation symbol intervals N_B which is needed to reach the upper bound on the minimum Euclidean distance [21].

TABLE 3.1

Punct. matrices for the investigated Punctured trellis coded CPFSK schemes, $h = 1/2$, 1REC.

Parent code	$G(D) = [1 + D + D^2; 1 + D^2]$		
Code Rate	$R = 1/2, p = 1$	$R = 2/3, p = 4$	$R = 3/4, p = 3$
P_{mat}	$(1,1)_o$	$(15,13)_o$	$(6,5)_o$
d_{min}^2	0.5	1.333	1.5
N_B	4	5	4
Parent code II	$G(D) = [1; D + D^2]$		
Code Rate	$R = 1/2, p = 1$	$R = 2/3, p = 4$	$R = 3/4, p = 3$
P_{mat}	$(1,1)_o$	$(15,13)_o$	$(6,5)_o$
d_{min}^2	6	0.6667	0.75
N_B	20	5	4

Different systems with puncture matrices given by Table 3.1 are simulated. For each channel SNR value, 100 source blocks with 8000 bits were used in the simulation. Fig. 3.3 shows the upper bound and the simulation results of BER performance for the investigated coded systems. It can be seen that the simulation results agree with the upper bound especially when the channel SNR increases. It also shows that the error probability performance for parent codes is better than the punctured codes. For example, for $G = [1; D + D^2]$, at BER of 10^{-4}, a performance gain of roughly ~ 4 dB can be obtained over the rate $3/4$ punctured coded system.

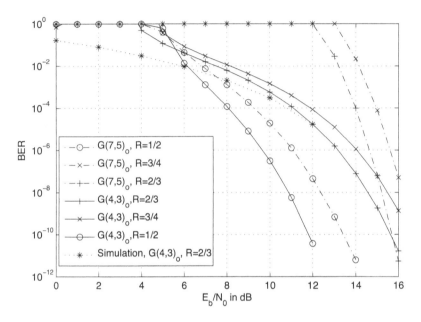

FIGURE 3.3: Analytical upper bounds and one simulation result for punctured trellis coded MSK (i.e., CPFSK with 1REC and $h = 1/2$) with $r = 1/2, 2/3, 3/4$, $G(D) = [1, D + D^2]$ and $G(D) = [1 + D^2; 1 + D + D^2]$.

3.5.2 SIMULATION RESULTS FOR INTERLEAVED DISTRIBUTED NETWORK CODED SYSTEM

Simulation for the interleaved distributed coded system with iterative decoding for a memoryless source over the AWGN channel is also performed. The encoding rates for the investigated systems are $r = \frac{2}{5}, \frac{1}{2}, \frac{2}{3}, \frac{3}{4}$. The generator polynomials of the distributed parent convolutional encoders are the same as those used for distributed network coded systems without interleaver. As we described earlier, a convergence analysis based on EXIT chart is mainly used to predict the decoding threshold of the systems with iterative decoding. The puncturing and repetition matrices, which are obtained by an exhaustive search from all of the possible puncturing and repetition patterns given a period for the one which gives the smallest decoding threshold, are given by P_{mat}^2 in the Table 3.2. The EXIT charts for the four studied systems are plotted in Fig. 3.4 for parent code $G = [1 + D + D^2; 1 + D^2]$. The decoding thresholds for the four studied distributed coded systems with iterative decoding are listed in Table 3.2.

TABLE 3.2

Puncturing and repetition matrices for the investigated interleaved distributed network coded systems.

Parent codes	$G = [1 + D + D^2; 1 + D^2]$			
Code Rate	$R = 2/5$ $p = 2$	$R = 1/2$ $p = 1$	$R = 2/3$ $p = 4$	$R = 3/4$ $p = 3$
P_{mat}^2	$Q_{2/5}$	$(1,1)_o$	$(15,13)_o$	$(5,3)_o$
SNR_{th}	0.8	1.1	2.8	4
$2BT_b$	2.35	1.88	1.41	1.25

This threshold is in dB of E_s/N_0. Also shown are 99% bandwidth $2BT_b$ for the investigated systems. Fig. 3.5 shows the simulation results of BER performance for the investigated distributed coded systems for parent code $G = [1+D+D^2; 1+D^2]$. It can be seen that the system with iterative approach significantly improves the BER performance compared with the one without iterative decoding.

3.6 SUMMARY

In this chapter, we proposed two distributed network coded modulation schemes for multiple access channel in a cellular system. We considered a system in which one source MT, cooperating with a cooperative MT, transmits information data packets to a BS via a relay node. In the cooperation phase, the source MT and the cooperative MT send the same data packet to the relay node using their own encoder via orthogonal channels. At the relay node, the received waveforms from both the MTs were symbol wise alternatively concatenated, then demodulated and decoded. There was an ARQ mechanism to ensure reliable transmissions from the MTs to the relay.

The correctly decoded codewords from the MTs were then fed into a network coded modulator to perform network encoding and digital phase coded modulation. In the first scheme which we investigated, the correctly decoded codewords from both the MTS were directly fed into the network coded modulator. While in the second scheme, the decoded codeword was first interleaved then fed into the network coded modulator. Analytical bound on BER performance for time varying trellis structure based distributed network coded modulation systems is derived. The bounds are shown asymptotically tight. These bounds can be served as the guideline for the design of the investigated distributed network coded modulation systems.

In the second scheme, we presented an iterative decoding approach for the interleaved distributed coded modulation systems. We used the EXIT chart to

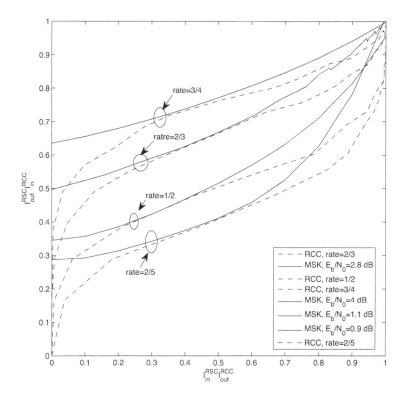

FIGURE 3.4: EXIT chart for interleaved distributed network coded MSK (i.e., CPFSK with 1REC and $h = 1/2$) with $r = 2/5, 1/2, 2/3, 3/4$, and $G(D) = [1 + D + D^2; 1 + D^2]$.

search the optimal puncturing and repetition patterns which can result in the lowest decoding threshold. The lower the decoding threshold is, the earlier the system converges. Compared with the first scheme, the second scheme gives much better performance due to the iterative approach. Since we considered rate compatible punctured and repetition distributed convolutional codes, the channel coding rate for each MT can be adaptive to the channel condition, therefore it was more robust to the channel in comparison with the fixed rate convolutional encoders.

Distributed Network Coded Modulation Schemes

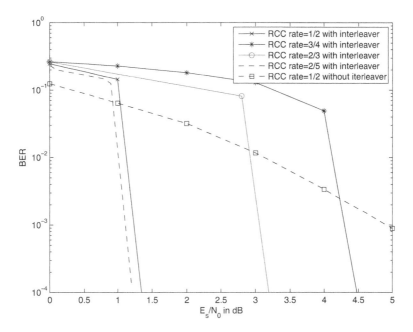

FIGURE 3.5: BER performance for interleaved distributed network coded MSK (i.e., CPFSK with 1REC and $h = 1/2$) with $r = 2/5, 1/2, 2/3, 3/4$, $G(D) = [1 + D + D^2; 1 + D^2]$.

1. A. Sendonaris, E. Erkip and B. Aazhang, "User cooperation diversity, part I: system description," *IEEE Trans. Commun.*, vol. 51, no. 11, pp. 1927–1938, Nov. 2003.

2. T. Hunter and A. Nosratinia, "Coopertion diversity through coding," in *Proc. IEEE ISIT'02*, 2002, pp. 220–221.

3. B. Zhao and M. C. Valenti, "Distributed turbo codes: towards the capacity of the relay channel," *IEEE VTC'03-Fall*, vol. 1, pp. 322–326, Oct. 2003.

4. S. Yiu, R. Schober and L. Lampe, "Distributed space-time block coding," *IEEE Trans. Commun.*, vol. 54, no. 7, pp. 1195–2006, Jul. 2006.

5. A. Chakrabarti, A. Baynast, A. Sabharwal and B. Aazhang, "Low density parity check codes for the relay channel," *IEEE JSAC*, vol. 25, no. 2, pp. 280–291, Feb. 2007.

6. R. Ahlswede, N. Cai, S. Y. R. Li, and R. W. Yeung, "Network Information Flow," *IEEE Trans. Inform. Theory*, vol. 46, no. 4, pp. 1204–1216, July 2000.

7. K. Doppler and M. Xiao, "Innovative concepts in peer-to-peer and network coding," *Tech. Rep. D1.3, CELTIC Telecommunication Solutions*, 2009.

8. P. Popovski and H. Yomo, "Physical network coding in two-way wireless relay channels," *IEEE Int. Conf. on Commun. (ICC)*, pp. 707–711, June 2007.

9. Z. Lin, *Rate Compatible Convolutional (RCC) Codes and Their Applications to Type II Hybrid ARQ Transmission*, Master Thesis, Chalmers University of Technology, Gothenburg, Sweden, Feb. 1998, http://www.ce.chalmers.se/TCT.

10. P. Frenger, P. Orten, T. Ottosson, and A. Svensson, "Multi-rate convolutional codes," *Tech. Rep. 21, Communication Systems Group, Departemnt of Signals and Systems*, Chalmers University of Technology, Apr. 1998.

11. Z. Lin and A. Svensson, "New rate compatible repetition convolutional codes," *IEEE Trans. Inform. Theory*, vol. 46, no. 7, pp. 2651–2659, Nov. 2000.

12. J. G. Proakis, *Digital Communication*, McGraw-Hill, New York, third edition, 1995.

13. A. J. Viterbi and L. K. Omura, *Principles of Digital Communication and Coding*, McGraw-Hill, New York, 1979.

14. S. Kallel and D. Haccoun, "Generalized type ii hybrid arq scheme using punctured convolutional coding," vol. 38, no. 11, pp. 1938–1946, Nov. 1990.

15. S. Lin and D. J. Costello, Jr., *Error Control Coding: Fundamentals and Applications*, Prentice Hall, New York, 1983.

16. B. Rimoldi, "A Decomposition Approach to CPM," *IEEE Trans. Inform. Theory*, vol. 34, no. 2, pp. 260–270, Mar. 1988.

17. J. M. Wozencraft and I. M. Jacobs, *Principles of Communication Engineering*, John Wiley & Sons, New York, 1965.

18. Z. Lin, *Joint Source-Channel Coding using Trellis Coded CPM*, Ph.D Thesis, Chalmers University of Technology, Gothenburg, Sweden, Jan. 2006, http://www.ce.chalmers.se/TCT.

19. E. Biglieri, "High-level modulation and coding for nonlinear satellite channels," *IEEE Trans. Commun.*, vol. COM-32, pp. 616–626, May 1984.

20. J. Shi and R. D. Wesel, "Efficient computation of trellis code generating function," *IEEE Trans. Commun.*, vol. 52, no. 2, pp. 219–227, Feb. 2004.

21. J. B. Anderson, T. Aulin, and C. E. Sundberg, *Digital Phase Modulation*, Plenum Press, New York, 1986.

22. S. Benedetto, D. Divsalar, G. Montorsi, and F. Pollara, "A Soft-Input Soft-Output APP Module for Iterative Decoding of Concatenated Codes," *IEEE Communications Letters*, vol. 1, no. 1, pp. 22–24, Jan. 1997.

23. L. R. Bahl, J. Cocke, F. Jelinek, and J. Raviv, "Optimal decoding of linear codes for minimizing symbol error rate," *IEEE Trans. Inform. Theory*, vol. IT-20, pp. 284–287, Mar. 1974.

24. S. ten Brink, "Convergence of iterative decoding," *Electron. lett.*, vol. 35, no. 10, pp. 806–808, May 1999.

25. J. Hagenauer, E. Offer, and L. Papke, "Iterative Decoding of Binary Block and Convolutional Codes," *IEEE Trans. Inform. Theory*, vol. 42, no. 2, pp. 429–445, Mar. 1996.

26. B. Scanavino, G. Montorsi, and S. Benedetto, "Convergence perperties of iterative decoders working at bit and symbol level," *Proc. IEEE Global Telecommunications Conference*, vol. 2, no. 25-29, pp. 1037–1041, Nov. 2001.

4 Lattice Network Coding for Multi-Way Relaying Systems

4.1 INTRODUCTION

High data rates and wide coverage are expected in future communications. Network coding [1] has received a lot of attention as a coding approach to improve the high data rate. The multi-way relay channel (MWRC) [2–4] is a notable network coding model in which all users use the same relay node, in which they exchange all of their information with no direct links between them. Satellite transmission is a common MWRC application scenario, in which all stations exchange information via satellite.

Multiple interpretations is an important concept in the MWRC. The relay node distributes coded packets that allow users to decode all of the other users' information, as shown in [5]. We discussed nested network coded system for multiple interpretations in Chapter 2. The basic concept is that different packets sent along a relay connection with independent linearly generators are merged, and then forwarded to various destinations. The authors in [6] propose a code design criterion to optimize the code profiles by combining convolutional and nested codes. The aforementioned coding schemes, on the other hand, are constrained by binary codes and only consider transmission in orthogonal channels. As a result, the spectrum efficiency of these schemes is low.

A lot of efforts have been done to build the nestled lattice codes in multi-user relay networks to improve the spectrum efficiency, such as Nazer and Gasper's compute-and-forward [7] proposal. Erez and Zamir demonstrate in [8] that nested lattice codes can achieve the AWGN channel capacity. The authors in [9] develops a general algebraic framework called lattice network coding based on PNC [10]. These papers propose concurrent transmissions from source to multi-user interference relay nodes. resulting in high spectrum efficiency. Due to the structure constraints of the code, these works are unable to achieve multiple interpretations with a single relay node.

In this chapter, we look at how to achieve multiple interpretations in MWRC with fading while maintaining high spectrum efficiency. In particular, we consider two time slots transmissions: in the first, all users transmit concurrently, and in the second, the network coded information is broadcast to all users by the relay. Over the finite field, we first present a novel nested convolutional lattice codes (NCLC) that can be interpreted by multiple sink nodes. The NCLC's codeword error rate (WER) is then given a theoretical upper bound. We also improve our NCLC by creating a code design criterion that reduces the derived WER. We construct a specific NCLC following the design criterion in simulations. The investigated code can realize multiple interpretations for each source. Simulation results verify the theoretical upper bound.

4.2 SYSTEM MODEL

4.2.1 SYSTEM MODEL

Consider a system model depicted in Fig. 4.1, where source nodes $[s_1, s_2, \cdots, s_L]$ wants to exchange information via a relay node r. The proposed scheme's simplified coding process can be broken down into the following steps: (1) All source nodes generate different information messages and encode them using linear generators which are mutually independent. (2) Each source node translates the encoded data into a lattice element. (3) In one time slot, all source nodes send their lattice codewords concurrently over wireless channels. (4) The superposition of all codewords is demodulated under power constraints at the relay. (5) In the second time slot, the relay node re-encodes the messages using lattice network coding and broadcasts the coded packet back to all source nodes. (6) Each source node can decode the other nodes' information based on the received coded packet from the relay and its own side information.

4.2.2 NESTED CONVOLUTIONAL CODES AND LATTICE NETWORK CODING

Consider the real field denoted by $\mathbb{R}$, the complex field denoted by $\mathbb{C}$, the integer field denoted by $\mathbb{Z}$, and the addition over the finite field denoted by $\oplus$. Furthermore, boldface lowercase and uppercase letters should be used to represent vectors and matrices, respectively.

Let $\mathbf{w}_\ell$ be the message generated by the source node s_ℓ independently and uniformly over the $\mathbb{F}_q$, where $\mathbb{F}_q$ represents a finite field of size q. Let $\mathbf{G}_\ell$ be the

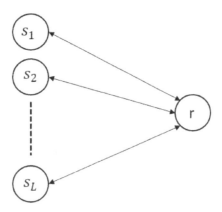

FIGURE 4.1: Multiple source nodes and single relay.

Lattice Network Coding for Multi-Way Relaying Systems 69

generator matrix over $\mathbb{F}_q$ at source s_ℓ, and denote the transpose of $\mathbf{G}$ by $\mathbf{G}^T$.

Similar to the nested convolutional codes over $GF(2)$ as mentioned in Chapter 3, the corresponding mathematical expression over $\mathbb{F}_q$ can be expressed as,

$$\mathbf{w}_1 \mathbf{G}_1 \oplus \mathbf{w}_2 \mathbf{G}_2 \oplus \cdots \oplus \mathbf{w}_L \mathbf{G}_L \\ = [\mathbf{w}_1, \mathbf{w}_2, \ldots, \mathbf{w}_L] [\mathbf{G}_1, \mathbf{G}_2, \ldots, \mathbf{G}_L]^T, \quad (4.1)$$

where $\mathbf{G}_i$ and $\mathbf{G}_j$ are independent $\forall i, j \in \{1, 2, \ldots, L\}, i \neq j$.

As we shall see, lattice codes provide an AWGN channel capacity with much more structure than random codes.

Denote by w a complex number such that the ring of integers $\mathbb{Z}[w] \triangleq \{a + bw | a, b \in \mathbb{Z}\}$ is a principle ideal domain (PID). In this chapter, a $\mathbb{Z}[w]$-lattice is defined as [9],

$$\Omega = \left\{ \boldsymbol{\omega} = \mathbf{G}_\Omega \mathbf{c} : \mathbf{c} \in \mathbb{Z}[i]^N \right\}, \quad (4.2)$$

where a lattice generator matrix $\mathbf{G}_\Omega \in \mathbb{C}^{N \times N}$ is full rank, N is the number of dimensions, and the set $\mathbb{Z}[i] \triangleq \{a + bi | a, b \in \mathbb{Z}\}$ are Gaussian integers.

A coarse lattice Ω' is a sublattice of a fine lattice Ω, i.e., $\Omega' \subset \Omega$. For the lattice network codes, the message space $\mathcal{W}$ is $\mathcal{W} = \Omega/\Omega'$, which can also be regarded as a module. The message rate for each user is defined as

$$R \triangleq \frac{1}{n} \log_2 |\mathcal{W}|. \quad (4.3)$$

A map $\mathcal{Q}_\Omega : \mathbb{C}^N \to \Omega$ is defined as a nearest-lattice-point (NLP) quantizer, which sends a point $\mathbf{x} \in \mathbb{C}^N$ to a nearest lattice point in Euclidean distance, i.e.,

$$\mathcal{Q}_\Omega(\mathbf{x}) \triangleq \arg \min_{\boldsymbol{\omega} \in \Omega} \| \mathbf{x} - \boldsymbol{\omega} \|. \quad (4.4)$$

Let $[\mathbf{s}] \bmod \Omega$ denote the quantization error of $\mathbf{s} \in \mathbb{C}^N$ with respect to the lattice Ω,

$$[\mathbf{s}] \bmod \Omega = \mathbf{s} - \mathcal{Q}_\Omega(\mathbf{s}). \quad (4.5)$$

Let $\mathcal{V}$ denote the fundamental Voronoi region of a lattice, which is the set of all points in $\mathbb{C}^N$ that are closest to the zero vector,

$$\mathcal{V} = \{\mathbf{s} : \mathcal{Q}_\Omega(\mathbf{s}) = 0\}. \quad (4.6)$$

Let $\psi(\mathbf{w})$ denote a map labeling the message to the points of Ω, similarly, $\psi^{-1}(\boldsymbol{\omega})$ signifies the inverse mapping process, i.e.,

$$\boldsymbol{\omega} = \psi(\mathbf{w}), \text{ and } \mathbf{w} = \psi^{-1}(\boldsymbol{\omega}), \quad (4.7)$$

where $\boldsymbol{\omega} \in \Omega$ and $\mathbf{w} \in \mathbb{F}_q$.

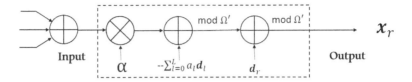

(a) The relay node's lattice network coding procedure.

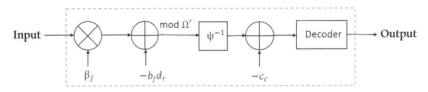

(b) The decoding procedure at the receiving node s_j.

FIGURE 4.2: NCLC procedure at the relay node and the source node s_j.

4.3 NESTED CONVOLUTIONAL LATTICE NETWORK CODES

The NCLC considers the nested convolutional codes' finite field $\mathbb{F}q$ as the lattice network codes' message space $\mathcal{W}$. As a result, the coded messages $\mathbf{w}_\ell \mathbf{G}_\ell$ are uniformly distributed on a field that is equivalent to the message space $\mathcal{W}$. Furthermore, we define a function F to represent an operation over a finite field, such as $F_q(\mathbf{w})$ for an operation over $\mathbb{F}_q$, and $F_{\Omega'}(\boldsymbol{\omega})$ for an operation over the fundamental Voronoi region of Ω'. Then we have,

$$F_q(\mathbf{w}) = [\mathbf{w}] \bmod q \, , \, F_{\Omega'}(\boldsymbol{\omega}) = [\boldsymbol{\omega}] \bmod \omega' , \\ \text{and } F_{\Omega'}(\boldsymbol{\omega}) = \psi(F_q(\mathbf{w})) \, , \, F_q(\mathbf{w}) = \psi^{-1}(F_{\Omega'}(\boldsymbol{\omega})) . \tag{4.8}$$

The transmitted signals are the points of the fine lattice ω,

$$\mathbf{x}_\ell = \psi(\mathbf{w}_\ell \mathbf{G}_\ell) . \tag{4.9}$$

Each transmitted signal is subject to an average power constraint given by,

$$\frac{1}{n} E\left[\| \mathbf{x}_\ell \|^2\right] \leq P . \tag{4.10}$$

We go through the NCLC coding method in greater depth in the following part. Each encoder is given a dither vector $\mathbf{d}$, which is created independently according to a uniform distribution over $\mathcal{V}(\Omega')$, the fundamental Voronoi region of Ω', for the sake of further proof. The source nodes' dither vectors $[\mathbf{d}_1, \mathbf{d}_2, \ldots, \mathbf{d}_L]$ are made available to the relay, and the relay's dither vector $\mathbf{d}_r$ is made available to each source node.

First, as shown in Fig. 4.2(a), the relay observes a channel output

$$\mathbf{y}_{sr} = \sum_{\ell=1}^{L} h_\ell \mathbf{x}_\ell + \mathbf{z}_{sr}, \qquad (4.11)$$

and the transmitted signal is

$$\mathbf{x}_\ell = [\mathbf{t}_\ell + \mathbf{d}_\ell] \bmod \Omega' = [\psi(\mathbf{w}_\ell \mathbf{G}_\ell) + \mathbf{d}_\ell] \bmod \Omega', \qquad (4.12)$$

where $\mathbf{t}_\ell$ denotes the coded messages on the fine lattice points.

With the developed lattice-partition-based compute-and-forward scheme of [7], we can obtain an optimal scheme at the relay node by choosing the scale factor α and the coefficient vector $\mathbf{a} \triangleq (a_1, a_2, \ldots, a_L)$, where $\alpha \in \mathbb{C}$, $\mathbf{a} \in \Omega$, then we have,

$$\begin{aligned}
\alpha \mathbf{y}_{sr} &= \sum_{\ell=1}^{L} \alpha h_\ell \mathbf{x}_\ell + \alpha \mathbf{z}_{sr} \\
&= \sum_{\ell=1}^{L} a_\ell \mathbf{x}_\ell + \underbrace{\sum_{\ell=1}^{L} (\alpha h_\ell - a_\ell) \mathbf{x}_\ell + \alpha \mathbf{z}_{sr}}_{\mathbf{n}} \\
&= \sum_{\ell=1}^{L} a_\ell \mathbf{x}_\ell + \mathbf{n},
\end{aligned} \qquad (4.13)$$

where h_ℓ is the fading coefficient for the link between source s_ℓ and relay node r, $\mathbf{z}$ is a vector of AWGN each of which follows a Gaussian distribution with zero mean and variance σ^2, and α can be obtained by [7],

$$\alpha = \frac{P_s \mathbf{h}^H \mathbf{a}}{P_s \parallel \mathbf{h} \parallel^2 + N_0}, \qquad (4.14)$$

where P_s is the source node transmission power, $\mathbf{h}^H$ represents the Hermitian transpose of $\mathbf{h}$ and N_0 is the average noise power.

We can remove dithers by using the following operations,

$$\begin{aligned}
\left[\alpha \mathbf{y}_{sr} - \sum_{\ell=1}^{L} a_\ell \mathbf{d}_\ell \right] \bmod \Omega' &= \left[\sum_{\ell=1}^{L} a_\ell \mathbf{x}_\ell + \mathbf{n} - \sum_{\ell=1}^{L} a_\ell \mathbf{d}_\ell \right] \bmod \Omega' \\
&= \left[\sum_{\ell=1}^{L} a_\ell \mathbf{t}_\ell + \mathbf{n} \right] \bmod \Omega',
\end{aligned} \qquad (4.15)$$

where $\sum_{\ell=1}^{L} a_\ell \mathbf{t}_\ell$ and $\mathbf{n}$ are independent each other.

Then there's the relay's lattice network coded packet, which should be sent back to each source node given the power constraints.

$$\begin{aligned}\mathbf{x}_r &= \left[\left[\sum_{\ell=1}^{L} a_\ell \mathbf{t}_\ell + \mathbf{n}\right] \bmod \Omega' + \mathbf{d}_r\right] \bmod \Omega' \\ &= \left[\sum_{\ell=1}^{L} a_\ell \mathbf{t}_\ell + \mathbf{n} + \mathbf{d}_r\right] \bmod \Omega'.\end{aligned} \qquad (4.16)$$

According to the reference [12], since $\sum_{\ell=1}^{L} a_\ell \mathbf{t}_\ell$ and $\mathbf{n}$ are independent, we can obtain that $\mathbf{x}_r$ and $\mathbf{n}$ are independent.

Next, as shown in Fig. 4.2(b), the source node s_j receives the lattice network coded packet from the relay node,

$$\begin{aligned}\mathbf{y}_{rs_j} &= h_j \mathbf{x}_r + \mathbf{z}_{rs_j} \\ &= h_j \left(\left[\sum_{\ell=1}^{L} a_\ell \mathbf{t}_\ell + \mathbf{n} + \mathbf{d}_r\right] \bmod \Omega'\right) + \mathbf{z}_{rs_j}.\end{aligned} \qquad (4.17)$$

Similarly, we remove the dithers at the jth receiver node by choosing some scalars β_j and b_j, where $\beta_j \in \mathbb{C}$, $b_j \in \Omega$,

$$\begin{aligned}&\left[\beta_j \mathbf{y}_{rs_j} - b_j \mathbf{d}_r\right] \bmod \Omega' \\ &= \left[\beta_j h_j \mathbf{x}_r + \beta_j \mathbf{z}_{rs_j} - b_j \mathbf{d}_r\right] \bmod \Omega' \\ &= \left[b_j \mathbf{x}_r + (\beta_j h_j - b_j)\mathbf{x}_r + \beta_j \mathbf{z}_{rs_j} - b_j \mathbf{d}_r\right] \bmod \Omega' \\ &= \left[b_j \sum_{\ell=1}^{L} a_\ell \mathbf{t}_\ell + \underbrace{b_j \mathbf{n} + (\beta_j h_j - b_j)\mathbf{x}_r + \beta_j \mathbf{z}_{rs_j}}_{\mathbf{m}_j}\right] \bmod \Omega' \\ &= \left[b_j \sum_{\ell=1}^{L} a_\ell \mathbf{t}_\ell + \mathbf{m}_j\right] \bmod \Omega',\end{aligned} \qquad (4.18)$$

where the optimal coefficient scalar b_j from the points of the fine lattice should be chosen to close to the channel coefficient according to [7].

Theorem 2. *Under the Minimum Mean Square Error (MMSE) decoding, the optimal scale factor β_j maximising the signal-to-effective-noise ratio (SENR) of Eq. (4.18) is,*

$$\beta_j = \frac{b_j P_r h_j}{P_r |h_j|^2 + N_0}, \qquad (4.19)$$

where P_r is the relay node transmission power.

Proof. First, we show that maximizing the SENR is equivalent to minimizing $\mathbf{m}_j/b_j$. From Eq. (4.18), we have

$$\left[b_j \sum_{\ell=1}^{L} a_\ell \mathbf{t}_\ell + \mathbf{m}_j \right] \mod \Omega'$$

$$= \left[b_j \left(\sum_{\ell=1}^{L} a_\ell \mathbf{t}_\ell + \frac{\mathbf{m}_j}{b_j} \right) \right] \mod \Omega' \quad (4.20)$$

$$= b_j \left(\sum_{\ell=1}^{L} a_\ell \mathbf{t}_\ell + \frac{\mathbf{m}_j}{b_j} \right) - \mathcal{Q}_{\Omega'}(\Theta),$$

where

$$\Theta = b_j \sum_{\ell=1}^{L} a_\ell \mathbf{t}_\ell + \mathbf{m}_j. \quad (4.21)$$

Because $\mathcal{Q}_{\Omega'}(\Theta)$ is the point on the coarse lattice Ω' and $\psi^{-1}(\mathcal{Q}_{\Omega'}(\Theta)) = 0$, we can regard $\mathcal{Q}_{\Omega'}(\Theta)$ as a regular shift of the signal. Hence, β_j should be chosen to minimize $\mathbf{m}_j/b_j$, which is equivalent to maximize the SENR.

We assume that each receiving node uses the MMSE detector. Let $f(\beta_j) = E\left[|\mathbf{m}_j/b_j|^2\right]$, from Eqs. (4.13) and (4.18), we have,

$$f(\beta_j) = \|\alpha \mathbf{h} - \mathbf{a}\|^2 P_s + |\alpha|^2 N_0$$
$$+ P_r \left|\frac{\beta_j}{b_j} h_\ell - 1\right|^2 + \left|\frac{\beta_j}{b_j}\right|^2 N_0. \quad (4.22)$$

It is apparent that $f(\beta_j)$ is convex. By deriving $f'(\beta_j) = 0$, we have,

$$\beta_j = \frac{b_j P_r h_\ell}{P_r |h_\ell|^2 + N_0}. \quad (4.23)$$

This completes the proof of Theorem 2. $\square$

Thus, we have the estimate of the desired linearly combination as

$$\hat{\mathbf{u}}_j = \psi^{-1}\left(\mathcal{Q}_\Omega\left(\left[b_j \sum_{\ell=1}^{L} a_\ell \mathbf{t}_\ell + \mathbf{m}_j\right] \mod \Omega'\right)\right)$$

$$= \psi^{-1}\left(\mathcal{Q}_\Omega\left(\left[F_{\Omega'}\left(b_j \sum_{\ell=1}^{L} a_\ell \mathbf{t}_\ell\right) + \mathbf{m}_j\right] \mod \Omega'\right)\right)$$

$$= \psi^{-1}\left(\mathcal{Q}_\Omega\left(F_{\Omega'}\left(b_j \sum_{\ell=1}^{L} a_\ell \mathbf{t}_\ell\right) + \mathbf{m}_j - \mathcal{Q}_{\Omega'}(\Theta)\right)\right) \quad (4.24)$$

$$= \psi^{-1}\left(F_{\Omega'}\left(b_j \sum_{\ell=1}^{L} a_\ell \psi(\mathbf{w}_\ell \mathbf{G}_\ell)\right) + \mathcal{Q}_\Omega(\mathbf{m}_j) - \mathcal{Q}_{\Omega'}(\Theta)\right)$$

$$= F_q\left(p_j \sum_{\ell=1}^{L} q_\ell \mathbf{w}_\ell \mathbf{G}_\ell\right) + \psi^{-1}(\mathcal{Q}_\Omega(\mathbf{m}_j))$$

$$= F_q(p_j \mathbf{W} \mathbf{Q} \mathbf{G}) + \psi^{-1}(\mathcal{Q}_\Omega(\mathbf{m}_j)),$$

where $\mathbf{W} = [\mathbf{w}_1, \mathbf{w}_2, \ldots, \mathbf{w}_L]$, $\mathbf{Q} = \mathrm{diag}\{q_1, q_2, \ldots, q_L\}$, $\mathbf{G} = [\mathbf{G}_1, \mathbf{G}_2, \ldots, \mathbf{G}_L]^T$, $p_j = \psi^{-1}([b_j] \bmod \Omega')$, $q_\ell = \psi^{-1}([a_\ell] \bmod \Omega')$, $p_j, q_\ell \in \mathbb{F}_q$, and

$$\Theta = \Gamma_{\Omega'}\left(b_j \sum_{\ell=1}^{L} a_\ell \mathbf{t}_\ell\right) + \mathbf{m}_j . \tag{4.25}$$

Note that the desired linear combination from the lattice network coded packet received at the j_{th} source node s_j can be formulated as

$$\mathbf{u}_j = \Gamma_q\left(p_j \sum_{\ell=1}^{L} q_\ell \mathbf{w}_\ell \mathbf{G}_\ell\right) \tag{4.26}$$
$$= \Gamma_q\left(p_j \mathbf{W}\mathbf{Q}\mathbf{G}\right) .$$

Therefore, it can be observed from Eqs. (4.24) and (4.26) that $\hat{\mathbf{u}}_j = \mathbf{u}_j$ if and only if $\psi^{-1}(\mathcal{Q}_\Omega(\mathbf{m}_j)) = 0$, or equivalently, $\mathcal{Q}_\Omega(\mathbf{m}_j) \in \Omega'$.

Furthermore, the decoding complexity can be reduced by canceling the prior known information. Denote by $\mathcal{K}_j$ the indices of the prior information known by the jth receiver. Denote by $\mathbf{c}_u$ the set of unknown information and let $\mathbf{c}_c$ be the set of prior known information. Then,

$$\mathbf{c}_u = \mathbf{u}_j \ominus \mathbf{c}_c$$
$$= \Gamma_q\left(p_j \sum_{\ell=1}^{L} q_\ell \mathbf{w}_\ell \mathbf{G}_\ell\right) \ominus \Gamma_q\left(p_j \sum_{\ell \in \mathcal{K}_j} q_\ell \mathbf{w}_\ell \mathbf{G}_\ell\right) \tag{4.27}$$
$$= \Gamma_q\left(p_j \sum_{\ell \notin \mathcal{K}_j} q_\ell \mathbf{w}_\ell \mathbf{G}_\ell\right) ,$$

where $\ominus$ is the subtraction operation over $\mathbb{F}_q$.

Given the assumption that all of the sources have prior knowledge of all of the assigned generators and can obtain a sequence of coefficients $\mathbf{q}$. Then, in two time slots, each receiver node can decode the unknown messages of all the other nodes and realize multiple interpretations.

4.4 PERFORMANCE ANALYSIS

In this section, we derive a theoretical upper bound for the NCLC's WER and develop a code design criterion that minimizes the derived WER to further optimize our NCLC.

First, we derive the error probability of $\hat{\mathbf{u}}_j$ at the j_{th} receiver node,

$$\Pr[\hat{\mathbf{u}}_j \neq \mathbf{u}_j] = \Pr[\mathcal{Q}_\Omega(\mathbf{m}_j) \notin \Omega'] . \tag{4.28}$$

With the reference [9], we have

$$\Pr[\mathcal{Q}_\Omega(\mathbf{m}_j) \notin \Omega'] \leq \Pr[\mathbf{m}_j \notin \mathcal{R}_\mathcal{V}(0)]$$
$$\leq \sum_{\boldsymbol{\omega} \in \mathrm{Nbr}(\Omega \setminus \Omega')} \exp\left(-\frac{v \|\boldsymbol{\omega}\|^2}{2}\right) E\left[\exp(v \mathrm{Re}\{\boldsymbol{\omega}^H \mathbf{m}_j\})\right], \forall v > 0$$
$$\tag{4.29}$$

Lattice Network Coding for Multi-Way Relaying Systems

where $\mathcal{R}_\mathcal{V}(\mathbf{0})$ denotes the Voronoi region of $\mathbf{0}$ in the set $\{\Omega \backslash \Omega'\} \cup \{\mathbf{0}\}$, and $\mathrm{Nbr}(\Omega \backslash \Omega')$ is the set of neighbors of $\mathbf{0}$ in $\{\Omega \backslash \Omega'\}$.

Subsequently, refer to Eqs. (4.13) and (4.18), we have

$$E\left[\exp(v\mathrm{Re}\{\boldsymbol{\omega}^H \mathbf{m}_j\})\right]$$
$$= E\left[\exp\left(v\mathrm{Re}\left\{\boldsymbol{\omega}^H\left(b_j \sum_{\ell=1}^{L}(\alpha h_\ell - a_\ell)\mathbf{x}_\ell + b_j \alpha \mathbf{z}_{sr} + (\beta_j h_j - b_j)\mathbf{x}_r + \beta_j \mathbf{z}_{rs_j}\right)\right\}\right)\right]$$
$$= E\left[\exp\left(v\mathrm{Re}\{\boldsymbol{\omega}^H(b_j \alpha \mathbf{z}_{sr} + (\beta_j h_j - b_j)\mathbf{x}_r + \beta_j \mathbf{z}_{rs_j})\}\right)\right]$$
$$\prod_{\ell} E\left[\exp(v\mathrm{Re}\{\boldsymbol{\omega}^H b_j(\alpha h_\ell - a_\ell)\mathbf{x}_\ell\})\right]$$
$$= \exp\left(\frac{1}{2}v^2 \parallel \boldsymbol{\omega} \parallel^2 \left(|b_j|^2|\alpha|^2 \sigma_{sr}^2 + P_r|\beta_j h_j - b_j|^2 + |\beta_j|^2 \sigma_{rs_j}^2\right)\right)$$
$$\prod_{\ell} E\left[\exp(v\mathrm{Re}\{\boldsymbol{\omega}^H b_j(\alpha h_\ell - a_\ell)\mathbf{x}_\ell\})\right]$$
$$= \exp\left(\frac{1}{4}v^2 \parallel \boldsymbol{\omega} \parallel^2 N_0 \left(|b_j|^2|\alpha|^2 + \frac{P_r}{N_0}|\beta_j h_j - b_j|^2 + |\beta_j|^2\right)\right)$$
$$\prod_{\ell} E\left[\exp(v\mathrm{Re}\{\boldsymbol{\omega}^H b_j(\alpha h_\ell - a_\ell)\mathbf{x}_\ell\})\right].$$

(4.30)

Here, we consider the lattice partition as a hypercube, refer to [9], we have

$$E\left[\exp(\mathrm{Re}\{\mathbf{v}^H \mathbf{x}\})\right] \leqslant \exp(\parallel \mathbf{v} \parallel^2 \delta^2/24), \quad (4.31)$$

where $\mathbf{x} \in \mathbb{C}^n$ is a complex random vector uniformly distributed over a hypercube $\delta \mathbf{U} \mathcal{H}_n$, $\delta > 0$ is a scalar factor, $\mathbf{U}$ is any $n \times n$ unitary matrix, and $\mathcal{H}_n$ is a unit hypercube in $\mathbb{C}^n$ defined by $\mathcal{H}_n = ([-1/2, 1/2] + i[-1/2, 1/2])^n$. Please note that for a hypercube $P = \frac{1}{n}E[\parallel \mathbf{x}_\ell \parallel^2] = \delta^2/6$. Thus, we have

$$E\left[\exp(v\mathrm{Re}\{\boldsymbol{\omega}^H \mathbf{m}_j\})\right]$$
$$= \exp\left(\frac{1}{4}v^2 \parallel \boldsymbol{\omega} \parallel^2 N_0\left(|b_j|^2|\alpha|^2 + \frac{P_r}{N_0}|\beta_j h_j - b_j|^2 + |\beta_j|^2\right)\right)$$
$$\prod_{\ell} E\left[\exp(v\mathrm{Re}\{\boldsymbol{\omega}^H b_j(\alpha h_\ell - a_\ell)\mathbf{x}_\ell\})\right]$$
$$\leq \exp\left(\frac{1}{4}v^2 \parallel \boldsymbol{\omega} \parallel^2 N_0\left(|b_j|^2|\alpha|^2 + \frac{P_r}{N_0}|\beta_j h_j - b_j|^2 + |\beta_j|^2\right)\right)$$
$$\prod_{\ell} \exp\left(\parallel v\boldsymbol{\omega} b_j(\alpha h_\ell - a_\ell) \parallel^2 P_s/4\right)$$
$$= \exp\left(\frac{1}{4}v^2 \parallel \boldsymbol{\omega} \parallel^2 N_0\left(|b_j|^2|\alpha|^2 + \frac{P_r}{N_0}|\beta_j h_j - b_j|^2 + |\beta_j|^2\right.\right.$$
$$\left.\left. + \frac{1}{4}v^2 \parallel \boldsymbol{\omega} \parallel^2 |b_j|^2 \parallel \alpha\mathbf{h} - \mathbf{a} \parallel^2 P_s\right)\right)$$
$$= \exp\left(\frac{1}{4}v^2 \parallel \boldsymbol{\omega} \parallel^2 N_0 Q(\alpha, \mathbf{a}, \beta_j, b_j)\right),$$

(4.32)

where the quantity $Q(\alpha, \mathbf{a}, \beta_j, b_j)$ is defined as

$$Q(\alpha, \mathbf{a}, \beta_j, b_j) = \frac{P_s}{N_0}|b_j|^2 \parallel \alpha \mathbf{h} - \mathbf{a} \parallel^2 + |b_j|^2|\alpha|^2 + \frac{P_r}{N_0}|\beta_j h_j - b_j|^2 + |\beta_j|^2. \quad (4.33)$$

Hence, by choosing $v = 1/\left(N_0 Q\left(\alpha, \mathbf{a}, \beta_j, b_j\right)\right)$, we obtain

$$\Pr[\widehat{\mathbf{u}}_j \neq \mathbf{u}_j] = \Pr[\mathcal{Q}_\Omega(\mathbf{m}_j) \notin \Omega']$$

$$\leqslant \sum_{\boldsymbol{\omega} \in \text{Nbr}(\Omega \setminus \Omega')} \exp\left(-\frac{v \parallel \boldsymbol{\omega} \parallel^2}{2} + \frac{1}{4}v^2 \parallel \boldsymbol{\omega} \parallel^2 N_0 Q(\alpha, \mathbf{a}, \beta_j, b_j)\right)$$

$$= \sum_{\boldsymbol{\omega} \in \text{Nbr}(\Omega \setminus \Omega')} \exp\left(-\frac{\parallel \boldsymbol{\omega} \parallel^2}{4 N_0 Q(\alpha, \mathbf{a}, \beta_j, b_j)}\right) \quad (4.34)$$

$$\approx K(\Omega/\Omega') \exp\left(-\frac{d^2(\Omega/\Omega')}{4 N_0 Q(\alpha, \mathbf{a}, \beta_j, b_j)}\right)$$

$$= K(\Omega/\Omega') \exp\left(-\frac{3 \text{SENR}_{\text{norm}} \gamma_c(\Omega/\Omega')}{2}\right),$$

where $\gamma_c(\Omega/\Omega')$, $d^2(\Omega/\Omega')$, and $K(\Omega/\Omega')$ denote the nominal coding gain, the squared minimum inter-coset distance, and the number of the nearest neighbors with $d^2(\Omega/\Omega')$ of the lattice partition Ω/Ω', respectively. We have SENR$_{\text{norm}}$ as [13]

$$\text{SENR}_{\text{norm}} = \frac{\text{SENR}}{2^R} = \frac{P}{2^R N_0 Q(\alpha, \mathbf{a}, \beta_j, b_j)}. \quad (4.35)$$

Let $V(\Omega)$ denote the volume of the Voronoi regions $\mathcal{V}(\Omega)$. The nominal coding gain can be expressed by [14]

$$\gamma_c(\Omega/\Omega') = \frac{d^2(\Omega/\Omega')}{V(\Omega)^{1/N}}, \quad (4.36)$$

where the number of dimensions N is equivalent to the packet length n.

We now calculate the WER of the jth node $\widehat{\mathbf{W}}_j$. As a result of [6, 15], we can obtain,

$$\Pr[\widehat{\mathbf{W}}_j \neq \mathbf{W}_j]$$

$$< \sum_{d=d_{free}^2}^{\infty} a_d (4 \Pr[\widehat{\mathbf{u}}_j \neq \mathbf{u}_j](1 - \Pr[\widehat{\mathbf{u}}_j \neq \mathbf{u}_j]))^{\sqrt{d}/2}, \quad (4.37)$$

where d_{free}^2 and a_d denote the minimum squared Euclidian distance and the number of paths at a squared Euclidian distance d from the all-zero path of the convolutional code corresponding to the "stacked" generator matrix $\mathbf{G}$ in (4.24), respectively.

The amount of error and packet length will not change because the cancellation process at all receiver nodes only discards the known information from the estimate of the desired linearly combined packets. As a result, the crossover probability remains at $\Pr[\widehat{\mathbf{u}}_j \neq \mathbf{u}_j]$ after the cancellation procedure has been completed. In other words, the cancellation process has no effect on the derived upper bound's performance.

Finally, we can have the following important design criterion from the Eqs. (4.33), (4.34), and (4.37).

1. The scalars α and β_j should be chosen such that $Q(\alpha, \mathbf{a}, \beta_j, b_j)$ is minimized;
2. The lattice partition Ω/Ω' should be designed such that $K(\Omega/\Omega')$ is minimized and $d^2(\Omega/\Omega')$ is maximized;
3. The "stacked" convolutional generator matrix should have the largest d_{free}^2 and the smallest a_d.

4.5 NUMERICAL SIMULATION RESULTS

In this section, we consider a MWRC with three source nodes and one relay node, and assign each source node a different linearly independent rate 1/3, memory 1 generator. We assume that all nodes have the same maximum transmission power P, that the lattice partition is $\mathcal{W} \cong \mathbb{Z}[i]/\delta\mathbb{Z}[i]$, where $\delta = 2 + 3i$, and that the finite field is $\mathbb{F}_{13}$, with size $q = 13$. Thus, $\mathcal{W} \cong \mathbb{F}_{13}$, the message rate is $R = \frac{1}{n}\log_2 13$, and the shaping is a rotated hypercube in $\mathbb{C}^N$.

We'll use the source node s_3 as an example node without losing generality. That is, we concentrate on realizing multiple interpretations at s_3 by extracting s_1 and s_2 information from the relay's lattice network coded packets. s_3 decodes the desired information by stacking the generate matrices of s_1 and s_2 as a rate 2/3 generate matrix, inversely mapping each received message from the relay to the corresponding non-binary convolutional codeword in $\mathbb{F}_{13}$, subtracting its own information based on the cancellation process shown in Eq. (4.27). We get a stacked generate matrix with the code rate 2/3 as follows based on the design criterion described in the previous section:

$$\begin{bmatrix} \mathbf{G}_1 \\ \mathbf{G}_2 \end{bmatrix} = \begin{bmatrix} 8+2D & 6+5D & 2+4D \\ 7+12D & 0 & 7D \end{bmatrix}, \quad (4.38)$$

$$d_{free}^2 = 9, \text{ and } a_{d|d=d_{free}^2} = 24. \quad (4.39)$$

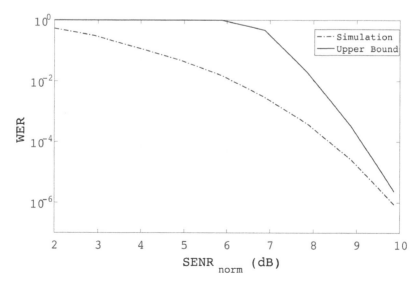

FIGURE 4.3: Upper bound on the WER and the simulation results for the investigated system.

Based on the code profile given in Eq. (4.38), we investigate the WER of s_3, i.e., $\Pr[\widehat{\mathbf{W}}_3 \neq \mathbf{W}_3]$. We have $\mathbf{W}_3 = [\mathbf{w}_1, \mathbf{w}_2]$, where $\mathbf{W}_3$ is the information matrix at the source node s_3, and is made up of information from the source nodes s_1 and s_2, respectively. The Monte-Carlo simulation result and the analytical upper bound of the WER are shown in Fig. 4.3. The simulation result and analytical upper bound are represented by the dotted and solid curves, respectively. As the SENR$_{\text{norm}}$ increases, we can see that the analytical upper bound gets closer to our Monte-Carlo simulation result.

4.6 CONCLUSION

We presented a nested convolutional coded lattice network coding scheme for multi-way relay channels with fading over a finite field in this chapter, which can achieve multiple interpretations. We began by describing the scheme's detailed coding process. We then developed a code design guideline by minimising the word error rate and derived a theoretical limit for the NCLC's WER. Finally, we built a specific NCLC based on the code design criterion in simulations. The NCLC can realize multiple interpretations for each source node. The upper bound on the WER has been validated by the simulation results.

1. R. Koetter and M. Medard, "An algebraic approach to network coding," *IEEE/ACM Transactions on Networking*, vol. 11, no. 5, pp. 782–795, Oct. 2003.

2. L. Ong, S. Johnson, and C. Kellett, "An optimal coding strategy for the binary multi-way relay channel," *IEEE Communications Letters*, vol. 14, no. 4, pp. 330–332, Apr. 2010.

3. L. Ong, C. Kellett, and S. Johnson, "Capacity theorems for the awgn multi-way relay channel," in *Proceedings of IEEE International Symposium on Information Theory (ISIT)*, Jun. 2010, pp. 664–668.

4. D. Gunduz, A. Yener, A. Goldsmith, and H. Poor, "The multi-way relay channel," in *Proceedings of IEEE International Symposium on Information Theory (ISIT)*, Jul. 2009, pp. 339–343.

5. L. Xiao, T. Fuja, J. Kliewer, and D. Costello, "Nested codes with multiple interpretations," in *Proceedings of 40th Annual Conference on Information Sciences and Systems (CISS)*, Mar. 2006, pp. 851–856.

6. Y. Ma, Z. Lin, H. Chen, and B. Vucetic, "Multiple interpretations for multi-source multi-destination wireless relay network coded systems," in *Proceedings of IEEE International Symposium on Personal, Indoor and Mobile Radio Communications (PIMRC)*, Sep. 2012.

7. B. Nazer and M. Gastpar, "Compute-and-forward: harnessing interference through structured codes," *IEEE Transactions on Information Theory*, vol. 57, no. 10, pp. 6463–6486, Oct. 2011.

8. U. Erez and R. Zamir, "Achieving 1/2 log (1+snr) on the awgn channel with lattice encoding and decoding," *IEEE Transactions on Information Theory*, vol. 50, no. 10, pp. 2293–2314, Oct. 2004.

9. C. Feng, D. Silva, and F. Kschischang, "An algebraic approach to physical-layer network coding," in *Proceedings of IEEE International Symposium on Information Theory (ISIT)*, Jun. 2010, pp. 1017–1021.

10. S. Zhang, S. C. Liew, and P. P. Lam, "Hot topic: physical-layer network coding," in *Proceedings of 12th Annual International Conference on Mobile Computing and Networking (MobiCom)*, 2006, pp. 358–365.

11. Z. Lin and B. Vucetic, "Power and rate adaptation for wireless network coding with opportunistic scheduling," in *Proceedings of IEEE International Symposium on Information Theory (ISIT)*, Jul. 2008, pp. 21–25.
12. B. Widrow and I. Kollá, *Quantization Noise*. Cambridge University Press, 2008.
13. Q. Sun and J. Yuan, "Lattice network codes based on eisenstein integers," in *Proceedings of 8th IEEE International Conference on Wireless and Mobile Computing, Networking and Communications (WiMob)*, Oct. 2012.
14. G. D. Forney, *MIT lecture nodes on Introdution to Lattice and Trellis Codes*.
15. T. Moon, *Error Correction Coding - Mathematical Methods and Algorithms*. Wiley-Interscience, 2005.

5 Nested LDGM-based Lattice Network Codes for Multi-Access Relaying Systems

5.1 INTRODUCTION

In Chapter 4, we introduced a class of nested convolutional lattice codes, which combine nested convolutional codes with lattice network coding to achieve high spectrum efficiency. The decoding complexity of NCLC, however, grows exponentially as the lattice dimensions increases. Since Low Density Parity Check (LDPC) codes can be decoded using the iterative decoding algorithm, constructing lattices based on such codes guarantees a manageable decoding complexity and can achieve more desirable performance. In this chapter, we consider Low Density Generator Matrix (LDGM) codes, a special type of LDPC codes [4] to construct lattices. Due to the sparseness of its generator matrix and the fact that the parity check matrix to/from generator matrix conversion is straightforward, the encoder complexity of LDGM codes is much less than that of LDPC codes.

Multiple interpretations is a critical concept in this work. [5], i.e. each user may decode the information provided by a relay node for all the others in a network encoded packet. With the purpose of multiple interpretations, the authors propose a novel nested code. The basic idea in [5] is to combine different packets with various linearly independent generators in a relay node and to transmit them to different destination nodes. A coding criterion for optimising the code profiles in [6] designs the codification codes jointly with the nested code. The above coding schemes, however, are limited by the binary codes, and only cover orthogonal transmission. There is, therefore, a relatively small spectrum efficiency in these schemes.

Much work has been done in the area of lattice codes, such as Nazer and Gasper's compute-and-forward proposal [7]. In [8], Erez and Zamir show that the additive white Gaussian noise (AWGN) may be achieved by nesting lattice code. A generic Algebraic Framework, dubbed network lattice coding in [9], is developed based on a network coding system of physical layer (PNC) [10]. These works present a notion of multi-user interference simultaneous transmission from source to relay nodes, which results in excellent spectrum efficiency. However, because to the restrictions of the code structures, these works cannot achieve MI with one relay node.

In this chapter, we construct a nested non-binary LDGM codes with lattice in a compute-and-forward scheme with fading to achieve multiple interpretation with low decoding complexity and high spectrum efficiency. We consider transmissions in two time slots, in which all users send their data to the relay at the same time (with multi-user interference) in the first time slot, and the relay forwards network coded data to the destination in the second time slot. We first propose a nested non-binary LDGM codes with lattice, which can achieve multiple interpretation at both relay and destination. Then we derive the detailed

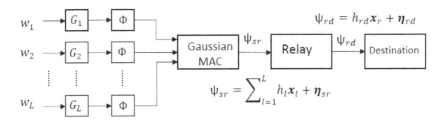

FIGURE 5.1: The compute-and-forward model.

coding process at both relay and destination nodes. At the destination node, we design a corresponding complexity reduced Lattice-based Extended Min-Sum (L-EMS) decoding algorithm. The performance of the low complexity decoding algorithm is quite good with respect to the important complexity reduction. Besides, we derive a theoretical upper bound for the codeword error rate (WER) of the proposed codes. Furthermore, we propose a Lattice-based Monte Carlo method to construct the optimized codes. In simulations, we construct a specific nested non-binary LDGM codes with lattice based on the proposed L-EMS decoder and Lattice-based Monte Carlo method. Simulation results show the performances of our L-EMS decoder and Monte Carlo results of the proposed codes.

The rest of this chapter is organized as follows. Section 5.2 for the introduced codes, and depicts the design of the nested structure of LDGM codes. In Section 5.3, binary LDGM codes with PNC as a special case. In Section 5.4, coding process of the proposed novel nested non-binary LDGM codes with lattice. The corresponding L-EMS decoding algorithm is demonstrated in Section 5.5. Section 5.6 shows the performance analysis of the proposed codes in the compute-and-forward scheme. Section 5.7 shows the code optimization using the proposed lattice-based Monte Carlo method. In Section 5.8, the numerical and simulation results are given and analyzed. Finally, Section 5.9 concludes this chapter.

5.2 SYSTEM MODEL

In this chapter, we consider a compute-and-forward scheme for a relay system with L transmitter, a single relay and a destination, where the relay receives messages from the L transmitters, forwards a lattice network coded packets to the destination, and the destination multiple interprets all the messages. We adopt the system model as in [7] and modified it with a destination, which is shown in Fig. 5.1. In the compute-and-forward scheme, each transmitter transmits an information block via a Gaussian multiple-access channel (MAC). Each information block belongs to the information space W, which is a finite module over a sub-ring R of $\mathbb{C}$. For the transmitter ℓ, it first encoded with itself generator matrix, then maps the codes message $\mathbf{w}_\ell \mathbf{G}_\ell \in W$ to an n-dimensional complex-valued signal $\mathbf{x}_\ell$ by the function $\phi: W \to \mathbb{C}^n$, and finally transmits it through the Gaussian MAC to the relay. Then the relay forwards the modulated signal to the destination. The detailed coding process will be presented in the following sections.

The mathematical operation of nested codes over $\mathbb{F}_q$ can be expressed by

$$\mathbf{w}_1 \mathbf{G}_1 \oplus \mathbf{w}_2 \mathbf{G}_2 \oplus \cdots \oplus \mathbf{w}_L \mathbf{G}_L$$
$$= [\mathbf{w}_1, \mathbf{w}_2, \ldots, \mathbf{w}_L][\mathbf{G}_1, \mathbf{G}_2, \ldots, \mathbf{G}_L]^T, \qquad (5.1)$$

where $\mathbf{G}_1, \mathbf{G}_2, \ldots, \mathbf{G}_L$ are mutually linearly independent.

Generally, a LDGM code can be structured as

$$\mathbf{G} = [\mathbf{I}_k, \mathbf{A}^T]$$
$$= \begin{bmatrix} \mathbf{I}_{k_1} & & & & \mathbf{A}_1^T \\ & \mathbf{I}_{k_2} & & & \mathbf{A}_2^T \\ & & \ddots & & \vdots \\ & & & \mathbf{I}_{k_L} & \mathbf{A}_L^T \end{bmatrix}. \quad (5.2)$$

Correspondingly, we can have different independent generators such as

$$\begin{aligned} \mathbf{G}_1 &= [\,\mathbf{I}_{k_1}\; \mathbf{0},\; \mathbf{A}_1^T\,] \\ \mathbf{G}_\ell &= [\,\mathbf{0}\; \mathbf{I}_{k_\ell}\; \mathbf{0},\; \mathbf{A}_\ell^T\,] \\ \mathbf{G}_L &= [\,\mathbf{0}\; \mathbf{I}_{k_L},\; \mathbf{A}_L^T\,]. \end{aligned} \quad (5.3)$$

Therefore, we have the linearly independent nested LDGM generators for each transmitter.

5.3 CODING PROCESS: NESTED BINARY LDGM CODES

Let s_ℓ denote the source nodes. Correspondingly, the generated binary information message and assigned generator from each source node are denoted by $\mathbf{w}_\ell$ and $\mathbf{G}_\ell$. We have the transmitted signal ζ_ℓ from each source node as,

$$\zeta_\ell = \phi(\mathbf{w}_\ell \mathbf{G}_\ell), \quad (5.4)$$

where ϕ denotes the modulation or mapping process.

Each transmitted signal is subject to an average power constraint given by,

$$\frac{1}{n} E\left[\|\zeta_\ell\|^2\right] \leq P. \quad (5.5)$$

Simultaneous transmission starting from each source node is assumed. Thus the observed signal from the relay node is

$$\psi_{sr} = \sum_{\ell=1}^{L} h_\ell \zeta_\ell + \eta_{sr}, \quad (5.6)$$

where h_ℓ is the channel coefficient of the link between source node s_ℓ and relay r, η represents the samples of AWGN with zero mean and variance σ^2.

Physical layer network coding (PNC) is assumed to be employed at the relay node to obtain the modulo 2 sum value of the received signal, we can have the log-likelihood ratio (LLR) of the module 2 sum value of the received signal, or we name it as the LLR of the XORed binary information, i.e., LLR $\left\{\bigoplus_{\ell=1}^{L} \mathbf{w}_\ell \mathbf{G}_\ell\right\}$.

For instance, we consider a simplified system model shown in Fig. 5.2, which consists of three source nodes, one relay node and one destination node. The relay can view a probability distribution shown in Fig. 5.3, where BPSK is assumed, and 0 is indicated by 1, 1 is indicated by −1.

The Gaussian distribution is defined by the formula

$$f(x) = \frac{1}{\sigma\sqrt{2\pi}} \exp\left\{-\frac{(x-\mu)^2}{2\sigma^2}\right\}, \quad (5.7)$$

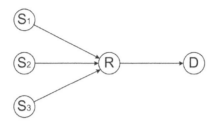

FIGURE 5.2: A special case with three sources, one relay, and one destination.

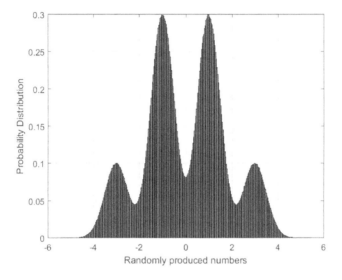

FIGURE 5.3: Probability distribution function of the received signal with three source nodes.

where the parameter μ is the mean or expectation of the distribution, and σ is the standard deviation.

Let $\mathbf{L} = \{1, 2, 3\}$ and $\ell \in \mathbf{L}$. We then have the derivation results shown as follows.

$$\begin{aligned}
&\text{LLR}\left\{\bigoplus_{\ell=1}^{L} \mathbf{w}_\ell \mathbf{G}_\ell\right\} \\
&= \ln \frac{\mathbb{P}\left\{\psi_{sr} \mid \bigoplus_{\ell=1}^{L} \mathbf{w}_\ell \mathbf{G}_\ell = 1\right\}}{\mathbb{P}\left\{\psi_{sr} \mid \bigoplus_{\ell=1}^{L} \mathbf{w}_\ell \mathbf{G}_\ell = 0\right\}} \\
&= \ln \frac{\mathbb{P}\left\{\psi_{sr} \mid \zeta_\mathbf{L} = -1\right\}\mathbb{P}\left\{\zeta_\mathbf{L} = -1\right\} + \sum_{\ell=1}^{L}\mathbb{P}\left\{\psi_{sr} \mid \zeta_\ell = -1, \zeta_{\mathbf{L}/\ell} = 1\right\}\mathbb{P}\left\{\zeta_\ell = -1, \zeta_{\mathbf{L}/\ell} = 1\right\}}{\mathbb{P}\left\{\psi_{sr} \mid \zeta_\mathbf{L} = 1\right\}\mathbb{P}\left\{\zeta_\mathbf{L} = 1\right\} + \sum_{\ell=1}^{L}\mathbb{P}\left\{\psi_{sr} \mid \zeta_\ell = 1, \zeta_{\mathbf{L}/\ell} = -1\right\}\mathbb{P}\left\{\zeta_\ell = 1, \zeta_{\mathbf{L}/\ell} = -1\right\}} \\
&= \ln \frac{\exp\left\{-\frac{(\psi_{sr}+h_\mathbf{L})^2}{2\sigma^2}\right\} + \sum_{\ell=1}^{L}\exp\left\{-\frac{(\psi_{sr}+h_\ell-h_{\mathbf{L}/\ell})^2}{2\sigma^2}\right\}}{\exp\left\{-\frac{(\psi_{sr}-h_\mathbf{L})^2}{2\sigma^2}\right\} + \sum_{\ell=1}^{L}\exp\left\{-\frac{(\psi_{sr}-h_\ell+h_{\mathbf{L}/\ell})^2}{2\sigma^2}\right\}},
\end{aligned}$$

(5.8)

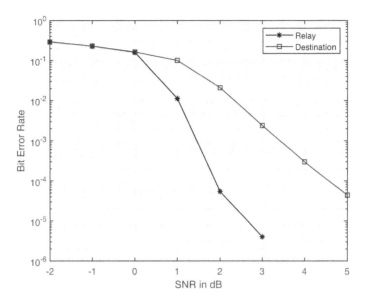

FIGURE 5.4: Code performance at relay and destination nodes with degree 6.

where,

$$\mathbb{P}\{\psi_{sr} \mid \zeta_\mathbf{L} = -1\} = \frac{1}{\sigma\sqrt{2\pi}} \exp\left\{-\frac{(\psi_{sr} + h_\mathbf{L})^2}{2\sigma^2}\right\},$$

$$\mathbb{P}\{\psi_{sr} \mid \zeta_\mathbf{L} = 1\} = \frac{1}{\sigma\sqrt{2\pi}} \exp\left\{-\frac{(\psi_{sr} - h_\mathbf{L})^2}{2\sigma^2}\right\},$$

$$\mathbb{P}\{\psi_{sr} \mid \zeta_\ell = -1, \zeta_{\mathbf{L}/\ell} = 1\} = \frac{1}{\sigma\sqrt{2\pi}} \exp\left\{-\frac{(\psi_{sr} + h_\ell - h_{\mathbf{L}/\ell})^2}{2\sigma^2}\right\},$$

$$\mathbb{P}\{\psi_{sr} \mid \zeta_\ell = 1, \zeta_{\mathbf{L}/\ell} = -1\} = \frac{1}{\sigma\sqrt{2\pi}} \exp\left\{-\frac{(\psi_{sr} - h_\ell + h_{\mathbf{L}/\ell})^2}{2\sigma^2}\right\},$$

$$\mathbb{P}\{\zeta_\mathbf{L} = -1\} = \mathbb{P}\{\zeta_\mathbf{L} = 1\} = \mathbb{P}\{\zeta_\ell = -1, \zeta_{\mathbf{L}/\ell} = 1\} = \mathbb{P}\{\zeta_\ell = 1, \zeta_{\mathbf{L}/\ell} = -1\} = \frac{1}{2^3}.$$
(5.9)

With the knowledge of the LLR $\left\{\bigoplus_{\ell=1}^{L} \mathbf{w}_\ell \mathbf{G}_\ell\right\}$, we have following two assumptions. First, we decode the LLR with LDGM code at the relay node. Second, we proceed hard-decision at the relay node and forward to the destination node.

The simulation results are shown in Fig. 5.4. A degree 6 size 1000×2000 LDGM code is employed. For the three source nodes, we employ the LDGM codes as divided,

$$\mathbf{G}^{1000 \times 2000} = \begin{bmatrix} \mathbf{G}_1^{200 \times 2000} \\ \mathbf{G}_2^{300 \times 2000} \\ \mathbf{G}_3^{500 \times 2000} \end{bmatrix}.$$

Because the same transmission power is assumed for all the nodes, the bit-error-rate for each source node with different code rate is consistent.

5.4 CODING PROCESS: NESTED NON-BINARY LDGM WITH LATTICE

In this section, we consider nested non-binary LDGM codes with lattice over finite field.

Let Ξ' be a coarse lattice, which is a sublattice of a fine lattice Ξ, i.e., $\Xi' \subset \Xi$. Denote the message space by $\mathcal{W}$, $\mathcal{W} = \Xi/\Xi'$, where Ξ/Ξ' denotes the set of all the cosets of Ξ' in Ξ.

Denote by $\mathbb{F}_q$ a finite field of size q, where q is a positive integer and $q > 2$. Let F be a function representing the operation over a finite field. and denote by $F_{\Xi'}(\boldsymbol{\xi})$ the operation over the fundamental Voronoi region of Ξ'. Then,

$$F_q(\mathbf{w}) = [\mathbf{w}] \bmod q \,, \; F_{\Xi'}(\boldsymbol{\xi}) = [\boldsymbol{\xi}] \bmod \Xi' \,,$$
$$\text{and } F_{\Xi'}(\boldsymbol{\xi}) = \phi\left(F_q(\mathbf{w})\right) \,, \; F_q(\mathbf{w}) = \phi^{-1}\left(F_{\Xi'}(\boldsymbol{\xi})\right), \tag{5.10}$$

where $\phi(\cdot)$ denotes a map labeling the message over $\mathbb{F}_q$ to the points over $\mathcal{V}(\Xi')$, the fundamental Voronoi region of Ξ', and $\phi^{-1}(\cdot)$ denotes the inverse process.

We show coding rules for the non-binary nested codes over $\mathbb{F}_q$. Denote by $\mathbf{w}_\ell$ the message generated by the ℓth source over $\mathbb{F}_q$ and let $\mathbf{G}_\ell$ be the ℓth source generator matrix over $\mathbb{F}_q$. Then the nested codes over $\mathbb{F}_q$ can be represented by

$$F_q\left(\sum_{\ell=1}^{L} \mathbf{w}_\ell \mathbf{G}_\ell\right) = F_q(\mathbf{W}\mathbf{G}) \tag{5.11}$$

where $\mathbf{W} = [\mathbf{w}_1, \mathbf{w}_2, \ldots, \mathbf{w}_L]$ and $\mathbf{G} = [\mathbf{G}_1, \mathbf{G}_2, \ldots, \mathbf{G}_L]^T$. Here $\mathbf{G}_1, \mathbf{G}_2, \ldots, \mathbf{G}_L$ are mutually independent.

For the nested non-binary LDGM Lattice codes, the finite field $\mathbb{F}_q$ of the codes can be treated as the message space $\mathcal{W}$ of the lattices. Thus, the coded message $\mathbf{w}_\ell \mathbf{G}_\ell$ is uniformly distributed over an equivalent field with the message space $\mathcal{W}$. We define the message rate for each source to be the same as $R \triangleq \frac{1}{n} \log_2 |\mathcal{W}| = r_\ell \log_2 q$, where r_ℓ is the LDGM code rate at teh ℓth source.

Let $\mathbf{d}$ be a dither vector over $\mathcal{V}(\Xi')$, the fundamental Voronoi region of Ξ'. The dither vectors $[\mathbf{d}_1, \mathbf{d}_2, \ldots, \mathbf{d}_L]$ at source nodes and the dither vector $\mathbf{d}_r$ at the relay are made available to each other.

Denote by $\mathbf{t}_\ell$ the coded messages on the fine lattice points, and ζ_ℓ the transmitted signal. Then,

$$\mathbf{t}_\ell = \phi(\mathbf{w}_\ell \mathbf{G}_\ell) \text{ and } \zeta_\ell = [\mathbf{t}_\ell + \mathbf{d}_\ell] \bmod \Xi' \,. \tag{5.12}$$

We assume an average power constraint is applied on each transmitted signal,

$$\frac{1}{n} E\left[\| \zeta_\ell \|^2\right] \leq P \,. \tag{5.13}$$

In the following part, we describe the coding process of the non-binary nested LDGM with lattice in more details.

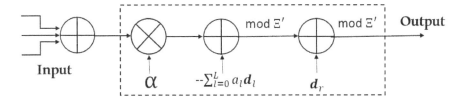

(a) The relay node's lattice network coding procedure.

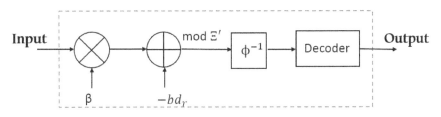

(b) The decoding procedure at the destination node.

FIGURE 5.5: Procedures of the non-binary nested LDGM with lattice at the relay and destination nodes.

First, as shown in Fig. 5.5(a), the relay observes a channel output

$$\psi_{sr} = \sum_{\ell=1}^{L} h_\ell \zeta_\ell + \eta_{sr}, \tag{5.14}$$

and the transmitted signal is

$$\begin{aligned} \zeta_\ell &= [\mathbf{t}_\ell + \mathbf{d}_\ell] \bmod \Xi' \\ &= [\phi(\mathbf{w}_\ell \mathbf{G}_\ell) + \mathbf{d}_\ell] \bmod \Xi'. \end{aligned} \tag{5.15}$$

Based on the lattice-partition-based compute-and-forward scheme [7], we can obtain an optimal scheme at the relay node by selecting the scale factor α, coefficient vector $\mathbf{a} \triangleq (a_1, a_2, \ldots, a_L)$, where $\alpha \in \mathbb{C}, \mathbf{a} \in \Xi$,

$$\begin{aligned} \alpha \psi_{sr} &= \sum_{\ell=1}^{L} \alpha h_\ell \zeta_\ell + \alpha \eta_{sr} \\ &= \sum_{\ell=1}^{L} a_\ell \zeta_\ell + \underbrace{\sum_{\ell=1}^{L} (\alpha h_\ell - a_\ell) \zeta_\ell + \alpha \eta_{sr}}_{\mathbf{n}} \\ &= \sum_{\ell=1}^{L} a_\ell \zeta_\ell + \mathbf{n}, \end{aligned} \tag{5.16}$$

where h_ℓ is the channel coefficient of the link between source node s_ℓ and relay r, η represents the samples of AWGN with zero mean and variance σ^2, and α is derived as [7],

$$\alpha = \frac{P_s \mathbf{h}^H \mathbf{a}}{P_s \parallel \mathbf{h} \parallel^2 + N_0}, \tag{5.17}$$

where P_s is the transmission power at each source node, $\mathbf{h}^H$ denotes the Hermitian transpose of $\mathbf{h}$ and N_0 is the average noise power.

To remove dithers under the power constraint, we have,

$$\begin{aligned}
&\left[\alpha\psi_{sr} - \sum_{\ell=1}^{L} a_\ell \mathbf{d}_\ell\right] \mod \Xi' \\
&= \left[\sum_{\ell=1}^{L} a_\ell \zeta_\ell + \mathbf{n} - \sum_{\ell=1}^{L} a_\ell \mathbf{d}_\ell\right] \mod \Xi' \\
&= \left[\sum_{\ell=1}^{L} a_\ell \mathbf{t}_\ell + \mathbf{n}\right] \mod \Xi',
\end{aligned} \tag{5.18}$$

where $\sum_{\ell=1}^{L} a_\ell \mathbf{t}_\ell$ and $\mathbf{n}$ are independent.

Then, at the relay, the generated lattice network coded packet can be expressed by,

$$\begin{aligned}
\zeta_r &= \left[\left[\sum_{\ell=1}^{L} a_\ell \mathbf{t}_\ell + \mathbf{n}\right] \mod \Xi' + \mathbf{d}_r\right] \mod \Xi' \\
&= \left[\sum_{\ell=1}^{L} a_\ell \mathbf{t}_\ell + \mathbf{n} + \mathbf{d}_r\right] \mod \Xi'.
\end{aligned} \tag{5.19}$$

According to the reference [12], since $\sum_{\ell=1}^{L} a_\ell \mathbf{t}_\ell$ and $\mathbf{n}$ are independent, we can obtain that ζ_r and $\mathbf{n}$ are independent.

Next, as shown in Fig. 5.5(b), the destination node d receives the lattice network coded packet from the relay node,

$$\begin{aligned}
\psi_{rd} &= h_{rd}\zeta_r + \eta_{rd} \\
&= h_{rd}\left(\left[\sum_{\ell=1}^{L} a_\ell \mathbf{t}_\ell + \mathbf{n} + \mathbf{d}_r\right] \mod \Xi'\right) + \eta_{rd}.
\end{aligned} \tag{5.20}$$

Similarly, we remove the dithers at the destination node by choosing some scalars β and b, where $\beta \in \mathbb{C}, b \in \Xi$,

$$\begin{aligned}
&[\beta\psi_{rd} - b\mathbf{d}_r] \mod \Xi' \\
&= [\beta h_{rd}\zeta_r + \beta\eta_{rd} - b\mathbf{d}_r] \mod \Xi' \\
&= [b\zeta_r + (\beta h_{rd} - b)\zeta_r + \beta\eta_{rd} - b\mathbf{d}_r] \mod \Xi' \\
&= \left[b\sum_{\ell=1}^{L} a_\ell \mathbf{t}_\ell + \underbrace{b\mathbf{n} + (\beta h_{rd} - b)\zeta_r + \beta\eta_{rs_j}}_{\mathbf{m}}\right] \mod \Xi' \\
&= \left[b\sum_{\ell=1}^{L} a_\ell \mathbf{t}_\ell + \mathbf{m}\right] \mod \Xi',
\end{aligned} \tag{5.21}$$

where b should be a value which is close to the channel coefficient [7].

Nested LDGM-based Lattice Network Codes for Multi-Access Relaying Systems 89

Theorem 3. *The optimal scale factor β that maximize the signal-to-effective-noise ratio (SENR) of Eq. (5.21), under the assumption of the Minimum Mean Square Error (MMSE) decoder, can be obtained by*

$$\beta = \frac{bP_r h_{rd}}{P_r |h_{rd}|^2 + N_0}, \qquad (5.22)$$

where P_r is the relay transmission power.

Proof. First, we show that maximizing the SENR is equivalent to minimizing $\mathbf{m}/b$. From Eq. (5.21), we have

$$\left[b \sum_{\ell=1}^{L} a_\ell \mathbf{t}_\ell + \mathbf{m}\right] \bmod \Xi'$$

$$= \left[b \left(\sum_{\ell=1}^{L} a_\ell \mathbf{t}_\ell + \frac{\mathbf{m}}{b}\right)\right] \bmod \Xi' \qquad (5.23)$$

$$= b \left(\sum_{\ell=1}^{L} a_\ell \mathbf{t}_\ell + \frac{\mathbf{m}}{b}\right) - \mathcal{Q}_{\Xi'}(\Theta),$$

where

$$\Theta = b \sum_{\ell=1}^{L} a_\ell \mathbf{t}_\ell + \mathbf{m}. \qquad (5.24)$$

Because $\mathcal{Q}_{\Xi'}(\Theta)$ is the point on the coarse lattice Ξ' and $\phi^{-1}(\mathcal{Q}_{\Xi'}(\Theta)) = 0$, we can regard $\mathcal{Q}_{\Xi'}(\Theta)$ as a regular shift of the signal. Hence, β should be chosen to minimize $\mathbf{m}/b$, which is equivalent to maximize the SENR.

It is assumed that the MMSE detector is employed at each receiver node. Let $f(\beta) = E\left[|\mathbf{m}/b|^2\right]$, from Eqs. (5.16) and (5.21), we have,

$$f(\beta) = \|\alpha \mathbf{h} - \mathbf{a}\|^2 P_s + |\alpha|^2 N_0$$
$$+ P_r \left|\frac{\beta}{b} h_{rd} - 1\right|^2 + \left|\frac{\beta}{b}\right|^2 N_0. \qquad (5.25)$$

It is apparent that $f(\beta)$ is convex. By deriving $f'(\beta) = 0$, we have,

$$\beta = \frac{bP_r h_{rd}}{P_r |h_{rd}|^2 + N_0}. \qquad (5.26)$$

This completes the proof of Theorem 3. $\square$

Thus, we have the estimate of the desired linearly combination as

$$\begin{aligned}
\hat{\mathbf{u}} &= \psi^{-1}\left(\mathcal{Q}_\Xi\left(\left[b\sum_{\ell=1}^{L}a_\ell\mathbf{t}_\ell+\mathbf{m}\right]\bmod \Xi'\right)\right)\\
&= \phi^{-1}\left(\mathcal{Q}_\Xi\left(\left[\Gamma_{\Xi'}\left(b\sum_{\ell=1}^{L}a_\ell\mathbf{t}_\ell\right)+\mathbf{m}\right]\bmod \Xi'\right)\right)\\
&= \phi^{-1}\left(\mathcal{Q}_\Xi\left(\Gamma_{\Xi'}\left(b\sum_{\ell=1}^{L}a_\ell\mathbf{t}_\ell\right)+\mathbf{m}-\mathcal{Q}_{\Xi'}(\Theta)\right)\right)\\
&= \phi^{-1}\left(\Gamma_{\Xi'}\left(b\sum_{\ell=1}^{L}a_\ell\phi\left(\mathbf{w}_\ell\mathbf{G}_\ell\right)\right)+\mathcal{Q}_\Xi(\mathbf{m})-\mathcal{Q}_{\Xi'}(\Theta)\right)\\
&= \Gamma_q\left(p\sum_{\ell=1}^{L}q_\ell\mathbf{w}_\ell\mathbf{G}_\ell\right)+\phi^{-1}\left(\mathcal{Q}_\Xi(\mathbf{m})\right)\\
&= \Gamma_q\left(p\mathbf{W}\mathbf{Q}\mathbf{G}\right)+\phi^{-1}\left(\mathcal{Q}_\Xi(\mathbf{m})\right),
\end{aligned}$$
(5.27)

where $\mathbf{W} = [\mathbf{w}_1, \mathbf{w}_2, \ldots, \mathbf{w}_L]$, $\mathbf{Q} = \text{diag}\{q_1, q_2, \ldots, q_L\}$, $\mathbf{G} = [\mathbf{G}_1, \mathbf{G}_2, \ldots, \mathbf{G}_L]^T$, $p = \phi^{-1}([b] \bmod \Xi')$, $q_\ell = \phi^{-1}([a_\ell] \bmod \Xi')$, $p, q_\ell \in \mathbb{F}_q$, and

$$\Theta = \Gamma_{\Xi'}\left(b\sum_{\ell=1}^{L}a_\ell\mathbf{t}_\ell\right)+\mathbf{m}.$$
(5.28)

Note that the desired linear combination from the lattice network coded packet received at the destination node can be formulated as

$$\begin{aligned}
\mathbf{u} &= \Gamma_q\left(p\sum_{\ell=1}^{L}q_\ell\mathbf{w}_\ell\mathbf{G}_\ell\right)\\
&= \Gamma_q\left(p\mathbf{W}\mathbf{Q}\mathbf{G}\right).
\end{aligned}$$
(5.29)

Therefore, it can be observed from Eqs. (5.27) and (5.30) that $\hat{\mathbf{u}} = \mathbf{u}$ if and only if $\phi^{-1}(\mathcal{Q}_\Xi(\mathbf{m})) = 0$, or equivalently, $\mathcal{Q}_\Xi(\mathbf{m}) \in \Xi'$.

Given the assumption that all the sources prior know all the assigned generators and can obtain a sequence of coefficients $\mathbf{q}$. Then based on the nested characteristic of the proposed codes, the destination node can decode all the messages from the source nodes.

5.5 L-EMS DECODING ALGORITHM

In this section, we present the ensemble complexity reduced decoding algorithm for the nested non-binary LDGM codes with lattice.

When a lattice point $\mathbf{x} \in \Xi$ is transmitted, at the destination node, the receiver decodes the desired message from the observed signal ψ_{rd}. A Lattice based Extended Min-Sum (L-EMS) algorithm is employed to the Tanner graph of Ξ to solve such a problem. A low-density Tanner graph companies a low iterative decoding complexity.

The L-EMS algorithm is performed by exchanging the truncated vector messages between variable nodes and check nodes through the edges of the Tanner graph in both directions iteratively. Each iteration consists of two steps, the variable node operation and the check node operation. With the use of the truncated messages, the EMS decoder [13] can efficiently reduce the order of complexity to $\mathcal{O}(n_m \log_2 n_m)$, with $n_m \ll q$. The L-EMS decoder keeps the feature of the efficient implementation of EMS and modifies it to be adaptive with lattice network coding.

A nested non-binary LDGM code is defined by a very sparse random parity check matrix H, whose components belong to a finite field $\mathbb{F}_q$. The matrix H consists of M rows and N columns, and the code rate is defined by $R \leq \frac{N-M}{N}$. We denote d_v the degree of a symbol nod and d_c the degree of a check node. A single parity check equation involving d_c variable nodes (codeword symbols) c_n is of the form:

$$\sum_{n=0}^{d_c-1} h_n c_n = 0 \text{ in } \mathbb{F}_q \tag{5.30}$$

where each h_n is a non-zero value of the parity matrix H.

We define the following notations. Let $\{\mathbf{V}_{cv_i}\}_{i \in \{0,\cdots,d_v-1\}}$ be the set of messages entering into a variable node v of degree d_v, and $\{\mathbf{U}_{v_ic}\}_{i \in \{0,\cdots,d_v-1\}}$ be the output messages for this variable node. Similarly, we define the messages $\{\mathbf{V}_{vc_i}\}_{i \in \{0,\cdots,d_c-1\}}$ (resp. $\{\mathbf{U}_{c_iv}\}_{i \in \{0,\cdots,d_c-1\}}$) at the input (reps. output) of a degree d_c check node.

The vector message $\mathbf{V}_{cv}$ and $\mathbf{U}_{vc}$ are limited to only n_m entries which are assumed to be the largest reliability values of the corresponding random variable. Moreover, the values in a message are sorted in decreasing order. We now present the steps of the L-EMS decoder that uses compensated-truncated message of size n_m.

1. Initialization: the n_m largest probabilities are obtain from the lattice-based channel output.
2. Variable-node update: the ranked output vector messages $\{\mathbf{U}_{v_ic}\}_{i \in \{0,\cdots,d_v-1\}}$ (of size n_m) associated to a variable node v passed to a check node c are computed given all the information propagated from all adjacent check nodes and the channel, except this check node itself.
3. Check-node update: for each check node, the ranked values $\{\mathbf{U}_{c_iv}\}_{i \in \{0,\cdots,d_c-1\}}$ sent from a check node to a variable node are defined as the the probabilities that the parity-check equation is satisfied if the variable node v is assumed to be equal to the truncated ranked message.

For steps 2) and 3), a recursive implementation combined with a forward/backward strategy is a well known efficient implementation of node update when the associated degree is larger than four. This implementation technique has been widely presented in the literature for non-binary LDPC codes.

5.6 PERFORMANCE ANALYSIS

In this section, we derive a theoretical upper bound for the WER of the codes.

First, we derive the error probability of $\widehat{\mathbf{u}}$ at the destination node d,

$$\Pr[\widehat{\mathbf{u}} \neq \mathbf{u}] = \Pr[\mathcal{Q}_{\Xi}(\mathbf{m}) \notin \Xi']. \tag{5.31}$$

With the reference [9], we have

$$\Pr[\mathcal{Q}_\Xi(\mathbf{m}) \notin \Xi'] \leqslant \Pr[\mathbf{m} \notin \mathcal{R}_\mathcal{V}(\mathbf{0})]$$
$$\leqslant \sum_{\boldsymbol{\xi} \in \text{Nbr}(\Xi \backslash \Xi')} \exp(-\frac{v \|\boldsymbol{\xi}\|^2}{2}) E\left[\exp(v\text{Re}\{\boldsymbol{\xi}^H \mathbf{m}\})\right], \forall v > 0 \tag{5.32}$$

where $\mathcal{R}_\mathcal{V}(\mathbf{0})$ denotes the Voronoi region of $\mathbf{0}$ in the set $\{\Xi\backslash\Xi'\}\cup\{\mathbf{0}\}$, and $\mathrm{Nbr}(\Xi\backslash\Xi')$ is the set of neighbors of $\mathbf{0}$ in $\{\Xi\backslash\Xi'\}$.

Subsequently, refer to Eqs. (5.16) and (5.21), we have

$$E\left[\exp(v\mathrm{Re}\{\boldsymbol{\xi}^H\mathbf{m}\})\right]$$
$$= E\left[\exp\left(v\mathrm{Re}\left\{\boldsymbol{\xi}^H\left(b\sum_{\ell=1}^L(\alpha h_\ell - a_\ell)\zeta_\ell + b\alpha\eta_{sr}\right.\right.\right.$$
$$+(\beta h_{rd} - b)\zeta_r + \beta\eta_{rd})\})]$$
$$= E\left[\exp\left(v\mathrm{Re}\{\boldsymbol{\xi}^H(b\alpha\eta_{sr} + (\beta h_{rd} - b)\zeta_r + \beta\eta_{rd})\}\right)\right]$$
$$\prod_\ell E\left[\exp(v\mathrm{Re}\{\boldsymbol{\xi}^H b(\alpha h_\ell - a_\ell)\zeta_\ell\})\right] \qquad (5.33)$$
$$= \exp\left(\frac{1}{2}v^2 \|\boldsymbol{\xi}\|^2 \left(|b|^2|\alpha|^2\sigma_{sr}^2 + P_r|\beta h_{rd} - b|^2 + |\beta|^2\sigma_{rd}^2\right)\right)$$
$$\prod_\ell E\left[\exp(v\mathrm{Re}\{\boldsymbol{\xi}^H b(\alpha h_\ell - a_\ell)\zeta_\ell\})\right]$$
$$= \exp\left(\frac{1}{4}v^2 \|\boldsymbol{\xi}\|^2 N_0 \left(|b|^2|\alpha|^2 + \frac{P_r}{N_0}|\beta h_{rd} - b|^2 + |\beta|^2\right)\right)$$
$$\prod_\ell E\left[\exp(v\mathrm{Re}\{\boldsymbol{\xi}^H b(\alpha h_\ell - a_\ell)\zeta_\ell\})\right].$$

Here, we consider the lattice partition as a hypercube, refer to [9], we have

$$E\left[\exp(\mathrm{Re}\{\mathbf{v}^H\zeta\})\right] \leqslant \exp(\|\mathbf{v}\|^2 \delta^2/24), \qquad (5.34)$$

where $\zeta \in \mathbb{C}^n$ is a complex random vector uniformly distributed over a hypercube $\delta\mathbf{U}\mathcal{H}_n$, $\delta > 0$ is a scalar factor, $\mathbf{U}$ is any $n \times n$ unitary matrix, and $\mathcal{H}_n$ is a unit hypercube in $\mathbb{C}^n$ defined by $\mathcal{H}_n = ([-1/2, 1/2] + i[-1/2, 1/2])^n$. Please note that for a hypercube $P = \frac{1}{n}E[\|\zeta_\ell\|^2] = \delta^2/6$. Thus, we have

$$E\left[\exp(v\mathrm{Re}\{\boldsymbol{\xi}^H\mathbf{m}\})\right]$$
$$= \exp\left(\frac{1}{4}v^2 \|\boldsymbol{\xi}\|^2 N_0 \left(|b|^2|\alpha|^2 + \frac{P_r}{N_0}|\beta h_{rd} - b|^2 + |\beta|^2\right)\right)$$
$$\prod_\ell E\left[\exp(v\mathrm{Re}\{\boldsymbol{\xi}^H b(\alpha h_\ell - a_\ell)\zeta_\ell\})\right]$$
$$\leq \exp\left(\frac{1}{4}v^2 \|\boldsymbol{\xi}\|^2 N_0 \left(|b|^2|\alpha|^2 + \frac{P_r}{N_0}|\beta h_{rd} - b|^2 + |\beta|^2\right)\right) \qquad (5.35)$$
$$\prod_\ell \exp\left(\|v\boldsymbol{\xi} b(\alpha h_\ell - a_\ell)\|^2 P_s/4\right)$$
$$= \exp\left(\frac{1}{4}v^2 \|\boldsymbol{\xi}\|^2 N_0 \left(|b|^2|\alpha|^2 + \frac{P_r}{N_0}|\beta h_{rd} - b|^2 + |\beta|^2\right)\right.$$
$$\left. + \frac{1}{4}v^2 \|\boldsymbol{\xi}\|^2 |b|^2 \|\alpha\mathbf{h} - \mathbf{a}\|^2 P_s\right)$$
$$= \exp\left(\frac{1}{4}v^2 \|\boldsymbol{\xi}\|^2 N_0 Q(\alpha, \mathbf{a}, \beta, b)\right),$$

where the quantity $Q(\alpha, \mathbf{a}, \beta, b)$ is defined as

$$Q(\alpha, \mathbf{a}, \beta, b)$$
$$= \frac{P_s}{N_0}|b|^2 \parallel \alpha \mathbf{h} - \mathbf{a} \parallel^2 + |b|^2|\alpha|^2 + \frac{P_r}{N_0}|\beta h_{rd} - b|^2 + |\beta|^2 . \quad (5.36)$$

Hence, by choosing $v = 1/(N_0 Q(\alpha, \mathbf{a}, \beta, b))$, we obtain

$$\Pr[\hat{\mathbf{u}} \neq \mathbf{u}] = \Pr[Q_\Xi(\mathbf{m}) \notin \Xi']$$
$$\leqslant \sum_{\boldsymbol{\xi} \in \mathrm{Nbr}(\Xi \setminus \Xi')} \exp\left(-\frac{v \parallel \boldsymbol{\xi} \parallel^2}{2} + \frac{1}{4}v^2 \parallel \boldsymbol{\xi} \parallel^2 N_0 Q(\alpha, \mathbf{a}, \beta, b)\right)$$
$$= \sum_{\boldsymbol{\xi} \in \mathrm{Nbr}(\Xi \setminus \Xi')} \exp\left(-\frac{\parallel \boldsymbol{\xi} \parallel^2}{4 N_0 Q(\alpha, \mathbf{a}, \beta, b)}\right) \quad (5.37)$$
$$\approx K(\Xi/\Xi') \exp\left(-\frac{d^2(\Xi/\Xi')}{4 N_0 Q(\alpha, \mathbf{a}, \beta, b)}\right)$$
$$= K(\Xi/\Xi') \exp\left(-\frac{3 \mathrm{SENR}_{\mathrm{norm}} \gamma_c(\Xi/\Xi')}{2}\right),$$

where $\gamma_c(\Xi/\Xi')$, $d^2(\Xi/\Xi')$, and $K(\Xi/\Xi')$ denote the nominal coding gain, the squared minimum inter-coset distance, and the number of the nearest neighbors with $d^2(\Xi/\Xi')$ of the lattice partition Ξ/Ξ', respectively. We have $\mathrm{SENR}_{\mathrm{norm}}$ as [14]

$$\mathrm{SENR}_{\mathrm{norm}} = \frac{\mathrm{SENR}}{2^R} = \frac{P}{2^R N_0 Q(\alpha, \mathbf{a}, \beta, b)}. \quad (5.38)$$

Let $V(\Xi)$ denote the volume of the Voronoi regions $\mathcal{V}(\Xi)$. The nominal coding gain can be expressed by [15]

$$\gamma_c(\Xi/\Xi') = \frac{d^2(\Xi/\Xi')}{V(\Xi)^{1/N}}, \quad (5.39)$$

where the number of dimensions N is equivalent to the packet length n.

5.7 CODE OPTIMIZATION USING LATTICE BASED MONTE CARLO METHOD

In this section, we will detail the lattice-based Monte Carlo method to optimize the codes. Since the structure of the nested non-binary LDGM codes is identical with that of a single link "stacked" non-binary LDGM code, thus we can facilitate the optimization of the proposed codes by optimizing a single link LDGM code with lattice.

As presented in the former section, the transmitted sequence ζ are the points on the lattice. Let ζ_i ($0 \leq i \leq N-1$) denote the ith signal. In the case of the AWGN channel, at the destination, the channel output corresponding to ζ_i is,

$$\psi_i = \zeta_i + \eta_i \mod \Xi'$$
$$= (\zeta_{i,rl} + j\zeta_{i,im}) + (\eta_{i,rl} + j\eta_{i,im}) \mod \Xi' \quad (5.40)$$
$$= \psi_{i,rl} + j\psi_{i,im} \mod \Xi',$$

where $z_{i,rl}$ and $z_{i,im}$ are realizations of independent Gaussian random variables with same variance $\sigma^2 = N_0/2$. Thus, the joint probability density of $\psi_{i,rl}, \psi_{i,im}$ in the case of lattice is

$$P(\psi_i|\zeta_i) = \frac{1}{2\pi\sigma^2} \sum_{k_1} \sum_{k_2} \exp\left(-\frac{(\psi_{i,rl} - \zeta_{i,rl} - qk_1)^2}{2\sigma^2}\right) \\ \times \exp\left(-\frac{(\psi_{i,im} - \zeta_{i,im} - qk_2)^2}{2\sigma^2}\right) \quad (5.41)$$

where q is the size of the finite field, as well as the number of lattice points in the fundamental Voronio region $\mathcal{V}(\Xi')$, and k_1, k_2 denote the ranges we need to compute.

According to the Bayes' theorem, we have

$$P(\zeta_i|\psi_i) = \frac{P(\psi_i|\zeta_i)P(\zeta_i)}{P(\psi_i)} \\ = t_i \sum_{k_1} \sum_{k_2} \exp\left(-\frac{(\psi_{i,rl} - \zeta_{i,rl} - qk_1)^2}{2\sigma^2}\right) \\ \times \exp\left(-\frac{(\psi_{i,im} - \zeta_{i,im} - qk_2)^2}{2\sigma^2}\right), \quad (5.42)$$

where $t_i = P(\zeta_i)/2\pi\sigma^2 P(\psi_i)$ is a constant if all the signals in $\mathcal{V}(\Xi')$ are transmitted with equal probability. The normalized constant t_i ensures $\sum_{\xi=1}^{q} P(\zeta_i = \zeta^{(\xi)}|\psi_i) = 1$, where $\zeta^{(\xi)}$ denotes the ξth symbol in $\mathcal{V}(\Xi')$. Let $\zeta^{(\xi)} = \zeta_{rl}^{(\xi)} + j\zeta_{im}^{(\xi)}$, we have the following a posterior probability (APP)

$$P(\zeta_i|\psi_i) = \frac{\sum_{k_1} \sum_{k_2} \exp\left(-((\psi_{i,rl} - \zeta_{i,rl} - qk_1)^2 + (\psi_{i,im} - \zeta_{i,im} - qk_2)^2)/2\sigma^2\right)}{\sum_{\xi=1}^{q} \sum_{k_1} \sum_{k_2} \exp\left(-((\psi_{i,rl} - \zeta_{i,rl}^{(\xi)} - qk_1)^2 + (\psi_{i,im} - \zeta_{i,im}^{(\xi)} - qk_2)^2)/2\sigma^2\right)}.$$

Proposition 1.

$$P(\zeta_i|\psi_i) = \frac{\sum_{k_1} \sum_{k_2} \exp\left(-\nu_1/2\sigma^2\right)}{\sum_{\xi=1}^{13} \sum_{k_1} \sum_{k_2} \exp\left(-\nu_2/2\sigma^2\right)}$$

where P_r is the transmission power at the relay, and

$$\nu_1 = (3\psi_{i,rl} - 2\psi_{i,im} - 3\zeta_{i,rl} + 2\zeta_{i,im} - 13k_1)^2 \\ + (2\psi_{i,rl} + 3\psi_{i,im} - 2\zeta_{i,rl} - 3\zeta_{i,im} - 13k_2)^2 \\ \nu_2 = (3\psi_{i,rl} - 2\psi_{i,im} - 3\zeta_{i,rl}^{(\xi)} + 2\zeta_{i,im}^{(\xi)} - 13k_1)^2 \\ + (2\psi_{i,rl} + 3\psi_{i,im} - 2\zeta_{i,rl}^{(\xi)} - 3\zeta_{i,im}^{(\xi)} - 13k_2)^2$$

Proof. We first construct a new coordinate system (ζ', ψ') as shown in Fig. 5.6. The lattice constellations are identical with Section 5.8.1. By algebraic operation, we have

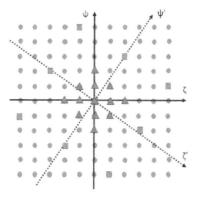

FIGURE 5.6: A new coordinate system for a practical Gaussian integer lattice constellations

$$\zeta' = \frac{1}{\sqrt{13}}(3\zeta - 2\psi)$$
$$\psi' = \frac{1}{\sqrt{13}}(2\zeta + 3\psi). \tag{5.43}$$

Recalling the APP equation shown in Section VII, and the two dimensions Gaussian function is constant for different coordinate systems. Consequently, we have the APP equation for the coordinate system (ζ', ψ') as

$$P(\zeta'_i|\psi'_i) = \frac{\sum_{k_1}\sum_{k_2} \exp\left(-((\psi'_{i,rl} - \zeta'_{i,rl} - qk_1)^2 + (\psi'_{i,im} - \zeta'_{i,im} - qk_2)^2)/2\sigma^2\right)}{\sum_{\xi=1}^{q}\sum_{k_1}\sum_{k_2} \exp\left(-((\psi'_{i,rl} - \zeta'^{(\xi)}_{i,rl} - qk_1)^2 + (\psi'_{i,im} - \zeta'^{(\xi)}_{i,im} - qk_2)^2)/2\sigma^2\right)}.$$

With Eq. 5.43, we can have

$$\psi'_{i,rl} = \frac{1}{\sqrt{13}}(3\psi_{i,rl} - 2\psi_{i,im})$$
$$\psi'_{i,im} = \frac{1}{\sqrt{13}}(2\psi_{i,rl} + 3\psi_{i,im})$$
$$\zeta'_{i,il} = \frac{1}{\sqrt{13}}(3\zeta_{i,rl} - 2\zeta_{i,im})$$
$$\zeta'_{i,im} = \frac{1}{\sqrt{13}}(2\zeta_{i,rl} + 3\zeta_{i,im}). \tag{5.44}$$

By inserting the these relations to the above APP equation, we then obtain the results of Proposition 1. □

In this work, we consider a practical lattice partition, which is chosen as a typical Gaussian integer $\mathcal{W} \cong \mathbb{Z}[i]/\delta\mathbb{Z}[i]$, where $\delta = 2 + 3i$, and the finite field is chosen as $\mathbb{F}_{13}$ where size $q = 13$. For this specific lattice, we derive the corresponding APP as given in *Proposition 1*.

Here, we consider the codes optimization based on a suboptimal LDGM code structure, because it is intractable to locate the optimal codes among a huge irregular LDGM matrix set. In particular, we consider only two parameters: average column weight $\overline{w}_c$ and the code rate $R = K/N$. The average row weight $\overline{w}_r$ can then be obtained by the following equation

$$\overline{w}_r = \overline{w}_c/(1-R) . \tag{5.45}$$

For a given $\overline{w}_c$, we specify the column profile that each column weight takes one or two values, either $\lfloor \overline{w}_c \rfloor$ or $\lfloor \overline{w}_c \rfloor + 1$. Mathematically, the column profile of the optimized codes is given by

$$c_j = \begin{cases} \lfloor \overline{w}_c \rfloor - \overline{w}_c + 1, & \text{if } j = \lfloor \overline{w}_c \rfloor ; \\ \overline{w}_c - \lfloor \overline{w}_c \rfloor, & \text{if } j = \lfloor \overline{w}_c \rfloor + 1 ; \\ 0, & \text{otherwise.} \end{cases} \tag{5.46}$$

where c_j denotes the proportion of columns with weight j in parity-check matrix $\mathbf{H}$, and $\lfloor \zeta \rfloor$ denotes the integer part of real number $\zeta > 0$.

Recall the definition in the Eq. (5.45), we can denote a certain nested non-binary LDGM code by parametric space $\Delta = (\overline{w}_c, R)$. The corresponding matrix set is denoted by $\mathcal{H}(\Delta) = \mathbf{H} : \Delta$. Each $\mathbf{H}$ in $\mathcal{H}(\Delta)$ is of dimension $N \times M$ where $M = N(1-R)$, and all the entries/operations are defined over $\mathbb{F}_q$. In this work, we endeavor to find the optimal value of $\overline{w}_c$, which can lead to a set of codes with good performance, and then employ PEG algorithm to obtain an optimal lattice-based nested non-binary LDGM code.

Let E_b be the average bit energy for transmission and N_0 be power of Gaussian noise, then we use $\gamma = E_b/N_0$ to denote the signal-to-noise ratio. In the rest of the chapters, we will use γ and E_b/N_0 to represent SNR interchangeably. For a given code rate R, we will optimize $\overline{w}_c$, with respect to SNR, under a certain bit error rate. Formally, we write

$$\overline{w}_{c,opt} = arg \min_{\overline{w}_c} \{\gamma(\overline{w}_c)\} . \tag{5.47}$$

As stated in [16], and widely adopted in the literature, the optimization in (5.47) must rely on simulation-based approaches.

As follows, we present the detailed steps of the lattice-based Monte Carlo method.

1. To initialize, input $(\overline{w}_c, R)$ and assign SNR_0 to E_b/N_0.
2. Randomly generate a sequence of K symbols, $\mathbf{w} = [w_0, w_1, \ldots, w_{K-1}]^T, (w \in \mathbb{F}_q)$, as uniform as possible.
3. Set $ii = 0$. For typically large code length, this set is not necessary, but for the medium or short code length, it is vital to test enough noise samples.
4. Generate a white Gaussian noise vector $\mathbf{n}_{ii}$, whose variance is controlled by E_b/N_0. The length of $\mathbf{n}_{ii}$ is defined by K/R.
5. Set $jj = 0$.
6. Randomly generate a parity check matrix of the nested non-binary LDGM codes: $\mathbf{H}_{jj} \in \mathcal{H}(\Delta), \Delta = (\overline{w}_c, R)$.
7. Generate the corresponding "stacked" generator matrix G_{jj}, and obtain the coded message $\mathbf{w}G_{jj}$. By lattice network coding, we have the transmitted signal on the lattice $\zeta_{jj} = \phi(\mathbf{w}G_{jj})$, and $\zeta = [\zeta_0, \zeta_1, \ldots, \zeta_{N-1}]^T$.
8. At the destination node, the channel output is observed as $\mathbf{y}_{(ii,jj)} = \zeta_{jj} + \mathbf{n}_{ii} \mod \Xi'$, and we can have the corresponding APP refer to (1). Then, by employing the lattice-based EMS algorithm discussed in Section 5.5, we can estimate Q_v^τ for each variable

node, where Q_v^τ is the probability of a estimated symbol equal to τ, and $1 \leq v \leq N$, $\tau \in \mathbb{F}_q$.
9. Repeat steps 3 to 8, so that the value of Q_v^τ can be converged. Then, we calculate the average entropy by

$$E = -\frac{1}{K} \sum_{v=0}^{K-1} \sum_{\tau=0}^{q-1} Q_v^\tau \log_q Q_v^\tau, \qquad (5.48)$$

where the range of v is shorten to K according to the structure of LDGM codes.
10. The value of $E = 0$ implies a complete removal of the ambiguity caused by the noisy channel, and thus, a complete retrieval of all the messages.

For different values of $\overline{w}_c$, we can use the algorithm above to attain the corresponding threshold SNRs. The optimal $\overline{w}_c$ is the one with minimum threshold SNR.

5.8 NUMERICAL AND SIMULATION RESULTS

In this section, we show the numerical and simulation results regarding to our proposed codes.

5.8.1 LATTICE SETTINGS

The lattice constellations are chosen as Fig. 5.7, where the grey round dots denote the fine lattice, red squares denote the coarse lattice, and the triangles denote the coding points. In this constellations figure, it is observed that the lattice partition is chosen as a typical Gaussian integer $\mathcal{W} \cong \mathbb{Z}[i]/\delta\mathbb{Z}[i]$, where $\delta = 2 + 3i$, and the finite field is chosen as $\mathbb{F}_{13}$ where size $q = 13$. Thus, $\mathcal{W} \cong \mathbb{F}_{13}$, the shaping is a rotated hypercube in $\mathbb{C}^N$, and the message rate is $R = \frac{1}{n} \log_2 13$.

5.8.2 LATTICE-BASED MONTE CARLO METHOD

The results for a certain rate code optimization using Lattice-based Monte Carlo method are given as Fig. 5.8. It shows the variation of the threshold SNR with different values of $\overline{w}_c$ in the design of proposed codes. The code length is assumed as $N = 2 \times 10^3$ and the code rate is $R = 0.5$.

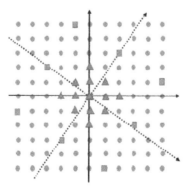

FIGURE 5.7: A practical Gaussian integer lattice constellations with the message space $\mathbb{F}_{13} \cong \mathcal{W}$, and $\mathcal{W} \cong \mathbb{Z}[i]/\delta\mathbb{Z}[i]$, where $\delta = 2 + 3i$

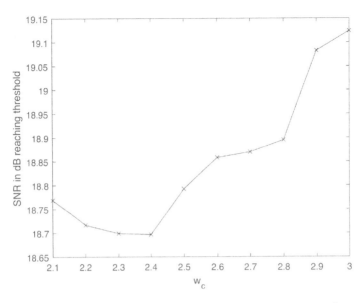

FIGURE 5.8: $\overline{w}_c$ optimization in the case of code length $N = 2 \times 10^3$, rate $R = 0.5$, and finite field $\mathbb{F}_{13}$.

Fig. 5.8 clearly demonstrates that there is an optimal average column weight for the proposed codes over the finite field $\mathbb{F}_{13}$. Specifically, we can have the optimal value $\overline{w}_c = 2.4$ in the case of $\mathbb{F}_{13}$. The existence of this trade-off in determining the optimal value of $\overline{w}_c$ for given R, can be understood as follows. Increasing $\overline{w}_c$ implies that we have more check equations for a single symbol and thus, the better the decoding performance. But refer to the Eq. (5.45), increasing $\overline{w}_c$ for a given code rate, on the other hand, also implies an increase in $\overline{w}_r$. Increasing $\overline{w}_r$ implies that more symbols are involved in a single check equation. Hence each symbol shares less information on average, which degrades the decoding performance. As a result, we have an optimal value of $\overline{w}_c$ for our proposed codes to achieve the threshold SNR.

Subsequently, with the optimal $\overline{w}_c$ and code rate R, and by employing PEG algorithm, we then can construct the parity-check matrix **H** and the corresponding "stacked" code generator matrix **G**.

5.8.3 PERFORMANCE FOR THE LATTICE-BASED EMS DECODER

We investigate the performances of the lattice-based EMS decoding algorithm for different n_m values over finite filed $\mathbb{F}_{13}$. The complexity of the proposed algorithm is dominated by $\mathcal{O}(n_m \log_2(n_m))$. In Fig. 5.9, the values of $n_m \in \{3, 5, 7, 13\}$ are compared over $\mathbb{F}_{13}$-AWGN channel. The parity check matrix **H** employed is the one optimized in Section 5.8.2. It can be observed that, with the increase of value of n_m, the performance of the proposed algorithm improves. When the value of $n_m = 7$, the corresponding performance appears almost identical with the most complex case $n_m = q = 13$, while the complexity is reduced by 2.44 times. Besides, it shows that the most dramatic improvement of the performance for

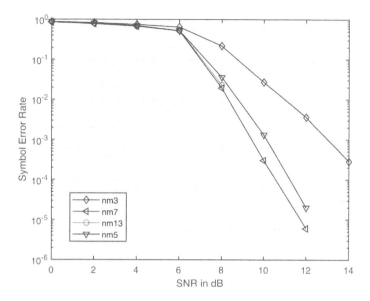

FIGURE 5.9: Comparison of Lattice-based EMS decoding performances for the different n_m values over finite filed $\mathbb{F}_{13}$.

the proposed algorithm starts from the value of $n_m = 3$ increasing to $n_m = 5$. Thus, we consider that $n_m = 5$ is a good complexity performance trade-off in our simulations.

5.8.4 PERFORMANCE OF THE NESTED NON-BINARY LDGM CODES WITH LATTICE

Now, we consider a practical case of our proposed system, which consists of three source nodes, one relay node, and one destination node. We assume that all the nodes have the same maximum transmission power P, the lattice partition is defined as 5.8.1.

The nested non-binary LDGM codes are constructed as given in 5.8.2. We then can assign different linearly independent LDGM codes for the different the source nodes. Specifically, we dived the "stacked" generator matrix as

$$\mathbf{G}^{1000 \times 2000} = \begin{bmatrix} \mathbf{G}_1^{200 \times 2000} \\ \mathbf{G}_2^{300 \times 2000} \\ \mathbf{G}_3^{500 \times 2000} \end{bmatrix}.$$

The decoder employs the L-EMS algorithm, the value of n_m equals to 5.

Next, we investigate the WER at the destination node, i.e., $\Pr[\widehat{\mathbf{W}} \neq \mathbf{W}]$, based on the code profile given above. We have $\mathbf{W} = [\mathbf{w}_1, \mathbf{w}_2, \mathbf{w}_3]$, where $\mathbf{W}$ is the information matrix at the destination node d, and is composed by the information $\mathbf{w}_1, \mathbf{w}_2, \mathbf{w}_3$ from the source nodes $s_1, s_2,$ and s_3, respectively. Fig. 5.10 illustrates the Monte-Carlo simulation result of the WER compared with $\text{SENR}_{\text{norm}}$ in dB.

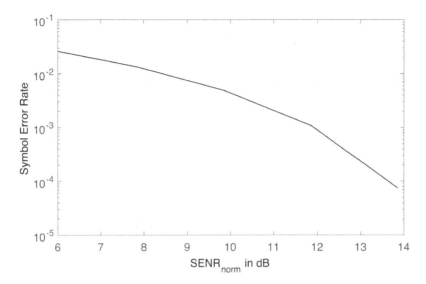

FIGURE 5.10: The Monte-Carlo simulation result of the WER of the message at the destination in the compute-and-forward scheme with three transmitters, one relay and one destination.

5.9 CONCLUSION

In this chapter, we proposed a novel nested non-binary LDGM codes with lattice to achieve multiple interpretations with low complexity decoding methods. Specifically, we constructed this novel codes in multi-way wireless relay network, derived the detailed coding process at both relay and destination nodes. At the destination node, we designed a corresponding complexity reduced Lattice-based Extended Min-Sum decoding algorithm. Besides, we proposed a Lattice-based Monte Carlo method to obtain a good nested non-binary LDGM codes. At last, simulation results show the good performances of our L-EMS decoder and Monte Carlo results of the proposed codes.

1. R. Koetter and M. Medard, "An algebraic approach to network coding," *IEEE/ACM Trans. Networking*, vol. 11, no. 5, pp. 782–795, Oct. 2003.
2. J. Li, J. Yuan, R. Malaney, M. Azmi, and M. Xiao, "Network coded ldpc code design for a multi-source relaying system," *IEEE Trans. Wireless Commun.*, vol. 10, no. 5, pp. 1538–1551, May. 2011.
3. J. H. Conway and N. J. A. Sloane, *Sphere Packings, Lattices, and Groups.* New York: Springer-Verlag, 1988.
4. J. F. Cheng and R. J. McEliece, "Some high-rate near capacity codecs for the gaussian channel," in *Proc. 34th Allerton Conference on Communication, Control and Computing*, 1996.
5. L. Xiao, T. Fuja, J. Kliewer, and D. Costello, "Nested codes with multiple interpretations," in *Proc. 40th Annual Conf. Information Sciences and Systems (CISS)*, Mar. 2006, pp. 851–856.

6. Y. Ma, Z. Lin, H. Chen, and B. Vucetic, "Multiple interpretations for multi-source multi-destination wireless relay network coded systems," in *Proc. Int. Symp. Personal, Indoor and Mobile Radio Communications (PIMRC)*, Sep. 2012.
7. B. Nazer and M. Gastpar, "Compute-and-forward: harnessing interference through structured codes," *IEEE Trans. Inform. Theory*, vol. 57, no. 10, pp. 6463–6486, Oct. 2011.
8. U. Erez and R. Zamir, "Achieving 1/2 log (1+snr) on the awgn channel with lattice encoding and decoding," *IEEE Trans. Inform. Theory*, vol. 50, no. 10, pp. 2293–2314, Oct. 2004.
9. C. Feng, D. Silva, and F. Kschischang, "An algebraic approach to physical-layer network coding," in *Proc. Int. Symp. Information Theory (ISIT)*, Jun. 2010, pp. 1017–1021.
10. S. Zhang, S. C. Liew, and P. P. Lam, "Hot topic: physical-layer network coding," in *Proc. 12th Annual Int. Conf. Mobile Computing and Networking (MobiCom)*, 2006, pp. 358–365.
11. Z. Lin and B. Vucetic, "Power and rate adaptation for wireless network coding with opportunistic scheduling," in *Proc. Int. Symp. Information Theory (ISIT)*, Jul. 2008, pp. 21–25.
12. B. Widrow and I. Kollá, *Quantization Noise*. Cambridge University Press, 2008.
13. A. Voicila, D. Declercq, F. Verdier, M. Fossorier, and P. Urard, "Low-complexity decoding for non-binary ldpc codes in high order fields," *IEEE Transactions on Communications*, vol. 58, no. 5, pp. 1365–1375, 2010.
14. Q. Sun and J. Yuan, "Lattice network codes based on eisenstein integers," in *Proc. 8th IEEE Int. Conf. Wireless and Mobile Computing, Networking and Communications (WiMob)*, Oct. 2012.
15. G. D. Forney, *MIT lecture nodes on Introduction to Lattice and Trellis Codes*.
16. B. Rong, T. Jiang, X. Li, and M. Soleymani, "Combine ldpc codes over gf(q) with q-ary modulations for bandwidth efficient transmission," *IEEE Transactions on Broadcasting*, vol. 54, no. 1, pp. 78–84, 2008.

6 Design of Soft Network Coding for Two-Way Relay Channels

6.1 INTRODUCTION

The recently emerging relay strategies have shown significant advantages for most future wireless applications, among which amplify-and-forward (AF) [1] and decode-and-forward (DF) [2] have been widely studied in the literature. Specifically, in terms of AF protocol, the relay amplifies the incoming signal without decoding it and then forwards it to the destination, which of course suffers from noise amplification at the relay. While DF protocol, in which the relay decodes the received signal, re-encodes it and then forwards it to the destination, propagates the erroneous decisions to the destination. Soft information forwarding (SIF) has been proposed to solve these kinds of problems [3]. One of the SIF based protocol is called estimate-and-forward (EF) and it performs better than the best of AF and DF in one-way relay channels [4]. Along with this, mutual information based forwarding (MIF) protocol has been proposed recently to further improve the error performance compared to EF protocol [5]. On the other hand, currently two-way relay channels (TWRC) have become increasingly appealing to both academics and industries, as network coding can be employed to achieve higher spectral efficiency. Quite a few relay strategies have been proposed in TWRC, such as [6–9].

In particular, the soft network coding scheme proposed in [9] has attracted a lot of attention. However, in [9] an assumption was made, that in the TWRC, the relay-to-source channels have limitless bandwidth, thus the soft information can be directly transmitted to either sources from the relay. In practice, due to the finite of channel bandwidth, we need to quantize the soft information. In this work, we are interested in combining joint trellis coded quantization (TCQ) and trellis coded modulation (TCM) over two way relay fading channels. TCQ functions as a feasible means here to quantize the soft information and shows modest computational complexity.

TCQ is proposed decades ago. In [10], it is proved that the quantization noise can be reduced without rate increase if a structured codebook with an expanded set of quantization levels is used. Although any modulation scheme can be used in conjunction with a TCQ, a TCQ/TCM system being consistent in bit and symbol rates, ensures that the squared distance between channel sequences is commensurate with the squared error in the quantization, which improves power-/bandwidth-efficiency and guarantees that the possible error occurred upon the channel symbol won't lead to much additional distortion in TCQ [11,12]. This joint TCQ/TCM system has been described explicitly in [12].

However in the two way relay fading channel, the soft information transmitted from the relay does not have the same probability distribution as described in [10]. Specifically, the probability distribution of the soft information varies with different signal-to-noise ratio (SNR) values of the source-to-relay channels. Based on large number of simulations with different channel SNRs, we found that the probability distribution of the soft information follows approximately Gaussian distribution in the low channel SNR region, while in the high channel SNR region, the values of soft information mostly become focusing around 1 and −1. We thus conjecture that under low channel SNR when the soft information follows approximately Gaussian distribution, in such a case, conventional TCQ can be used in the way described in [10], and under high channel SNR values, conventional TCQ cannot be used directly to quantize the soft information. In such a case, we can either use the DF scheme or modify the conventional TCQ scheme to adapt the situation where the soft information values concentrate on 1 or −1. The reason why we consider DF scheme is because under high channel SNR, a few or even no decoding errors will occur at the relay, thus DF scheme can be employed. In this work, based on the network coded EF protocol proposed in [9], we use EF with conventional TCQ scheme to transmit the soft information under low source-to-relay channel SNR, and employ two different soft information forwarding schemes at the relay under high source-to-relay channel SNR. One is to use DF scheme, the other is still using EF with TCQ scheme but the preset codebook for TCQ is different from teh conventional TCQ, and we call such a scheme as the refined EF with TCQ scheme. According to the simulation, the proposed schemes outperform both DF and AF protocols over two way relay fading channels.

The rest of this chapter is organized as follows. In Section 6.2, system model of the TWRC system is presented, the EF with TCQ scheme is also introduced. In Section 6.3, we propose two soft information delivery schemes under high source-to-relay channel SNR. Performance analysis for the two soft information forwarding schemes over two way relay fading channels is presented in Section 6.4. Simulation results are shown in Section 6.5, and conclusion is drawn in Section 6.6.

6.2 SYSTEM MODEL

As for TWRC shown in Fig. 6.1, where the two sources transmit or receive information at different time slots. At the first time slot, S_1 transmits its own signals to both the relay and S_2. At the second time slot, S_2 transmits its own signals to both the relay and S_1. The received signal at the relay from either source and the received signal at the other source from the direct link can be respectively calculated by

$$r_{S_i R} = \sqrt{E_S} h_{S_i R} x_{S_i} + n_{S_i R},$$

$$r_{S_i S_j} = \sqrt{E_S} h_{S_i S_j} x_{S_i} + n_{S_i S_j}, \quad (6.1)$$

where x_{S_i} represents the signal transmitted from either of the two sources S_i, where $i \in \{1, 2\}$, E_S is defined as the transmission power at either source, $h_{S_i S_j}$ and $h_{S_i R}$ represent the channel coefficient between two sources, and the channel coefficient between either source S_i and relay, respectively. $n_{S_i S_j}$ and $n_{S_i R}$ are the noise samples at one source and the relay, which are both defined as Gaussian distributed random variables with zero mean value and the variance σ^2. $r_{S_i R}$ and $r_{S_i S_j}$ represent the signal received at the relay and at the source from the other source. The SNR of the source-to-relay channel can be defined by $\rho_{S_i R} \triangleq h^2_{S_i R} E_S / \sigma^2$.

At the third time slot, the relay processes the received signals and then broadcasts the processed signal to both S_1 and S_2. In this stage, we apply TCQ to quantize the soft network coded information defined in [9]. At first, the network coded symbol is defined as $x_R \triangleq x_{S_1} x_{S_2}$. Afterwards minimum mean squared error (MMSE) estimation of x_R is calculated, namely the expectation of $x_{S_1} x_{S_2}$, $\mathrm{E}\left(x_{S_1} x_{S_2} | r_{S_1 R}, r_{S_2 R}\right)$. Due to the independence of the

Design of Soft Network Coding for Two-Way Relay Channels

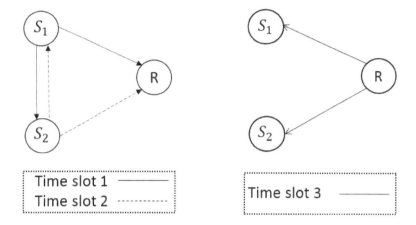

FIGURE 6.1: A two-way relay network.

received r_{S_1R} and r_{S_2R}, the MMSE estimation at the relay is given by

$$E(x_{S_1}x_{S_2}|r_{S_1R}, r_{S_2R}) = E(x_{S_1}|r_{S_1R}) E(x_{S_2}|r_{S_2R})$$

$$= \tanh\left(\frac{LLR_{x_{S_1},R}}{2}\right) \tanh\left(\frac{LLR_{x_{S_2},R}}{2}\right), \quad (6.2)$$

where $LLR_{x_{S_i},R}$ represents the log-likelihood ratio (LLR) of r_{S_iR} at the relay, which is calculated as $LLR_{x_{S_i},R} = \ln\frac{p(x_{S_i}=1|h_{S_iR}, r_{S_iR})}{p(x_{S_i}=-1|h_{S_iR}, r_{S_iR})} = \frac{2\sqrt{E_s}h_{S_iR}r_{S_iR}}{\sigma^2}$.

We denote by $f(r_{S_1R}, r_{S_2R})$ the power-normalized relay function in EF protocol, which is given by

$$f(r_{S_1R}, r_{S_2R}) = \frac{\tanh\left(\frac{LLR_{x_{S_1},R}}{2}\right) \tanh\left(\frac{LLR_{x_{S_2},R}}{2}\right)}{\sqrt{E\left[\left|\tanh\left(\frac{LLR_{x_{S_1},R}}{2}\right) \tanh\left(\frac{LLR_{x_{S_2},R}}{2}\right)\right|^2\right]}}, \quad (6.3)$$

We apply TCQ to quantize the power-normalized relay function $f(r_{S_1R}, r_{S_2R})$, and denote $\hat{f}^*(r_{S_1R}, r_{S_2R})$ as the quantized relay function. In this way, the signal received by either source from the relay can be given by $r_{RS_i} = \hat{f}^*(r_{S_1R}, r_{S_2R})$, which can be equivalently viewed as

$$r_{RS_i} = f(r_{S_1R}, r_{S_2R}) + n_{equiv\_RS_i}, . \quad (6.4)$$

Here $\hat{f}^*(r_{S_1R}, r_{S_2R})$ represents the received quantized signal at the receiver, $n_{equiv\_RS_i}$ is regarded as the quantization noise.

FIGURE 6.2: A two-way relay network.

The system model is shown in Fig. 6.2, in which a joint TCQ/TCM system is followed after the power-normalized relay function is calculated.

In the joint TCQ/TCM system, we assume that the encoding rate for TCQ is R b/sample, and for TCM the transmission rate is R b/symbol. They both use the identical trellis to ensure the consistent labeling in the trellis diagram, which can guarantee that likely channel error events will only lead to small additional TCQ distortion [12]. TCQ and TCM both use the Viterbi algorithm, which is used to find the appropriate sequence path of quantization levels which has the shortest Euclidean distance to the corresponding input. Let $\mathbf{x}$ be the input sequence of the source encoder with length m and $\hat{\mathbf{x}}$ be the corresponding output sequence, then the Euclidean distance between the two sequences is given by

$$\mathrm{d}\left(\mathbf{x}, \hat{\mathbf{x}}\right) = \sqrt{\sum_{i=1}^{m}(x_i - \hat{x}_i)^2}. \tag{6.5}$$

In TCQ, the Viterbi algorithm is used to find the output sequence which minimizes $\mathrm{d}\left(\mathbf{x}, \hat{\mathbf{x}}\right)$. Equivalently it can be viewed as minimizing the mean squared error (MSE),

$$\rho_m\left(\mathbf{x}, \hat{\mathbf{x}}\right) = \mathrm{d}^2\left(\mathbf{x}, \hat{\mathbf{x}}\right)/m. \tag{6.6}$$

Usually TCQ uses a rate $R/(R+1)$ convolutional encoder to define the structure of trellis, and the Viterbi algorithm is used to search all possible paths in the trellis and selects the minimum distortion path. For each incoming sample, a rate R b/sample TCQ maps it into one of 2^{R+1} reproduction levels (codewords), the set of the reproduction levels is regarded as a codebook. Fig. 6.2 shows a joint TCQ/TCM system model. The trellis coded quantizer generates a sequence consisting of the 2^{R+1} reproduction levels from the defined codebook, these levels are directly mapped to the symbols made up of the 2^{R+1}-point TCM alphabet [10–12].

At the receiver, the decoding is accomplished by first using the Viterbi algorithm to find the path which has the MMSE to the received signals, and then mapping the selected path back into the TCQ levels.

6.3 TCQ CODEBOOK DESIGN

In order to compete with DF, which only requires 1 b/sample after hard decision is made at the relay, we consider to use a rate of 1 b/sample TCQ to quantize the soft information. We assume that channels between all terminals experience independent slow Rayleigh fading, that is, the channel coefficients for all the channels are constant during one data block but change independently from block to block. Due to slow fading, it is possible for receivers to accurately estimate the channel state information. Hence, we will assume perfect channel-state-information (CSI) at all the receivers. Besides, the codebook for TCQ should be synchronized at both the relay and the receiver at each source. This can be achieved by presetting different codebooks for TCQ according to different source-to-relay channel SNRs.

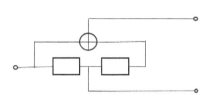 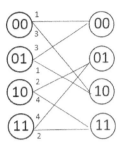

FIGURE 6.3: A $(5,2)_8$ convolutional encoder and its trellis diagram.

Source-to-relay CSI is known by the relay, and destination also knows the source-to-channel CSI, thus the synchronization can be achieved.

In simulation, Lloyd-Max quantizer is used to find the corresponding codebook for each m-symbol block. However, in (6.3), it is shown that $f(r_{S_1R}, r_{S_2R})$ approaches 1 or -1 when the LLR values of the received symbols are large enough. In the case that all the values of $f(r_{S_1R}, r_{S_2R})$ in one block are either 1 or -1, the output levels of Lloyd-Max quantizer might be $[-1 - 1 - 11]$. With this codebook for TCQ, quite a few mapping errors might occur. This can be demonstrated as follows. With a rate-1/2 convolutional encoder, for an integral rate 1 b/sample encoding, 2^2 reproduction levels (codewords) are used. As long as the structure of the convolutional encoder is set, the branch labeling in the trellis is fixed. And the mapping principle is given in (6.5), TCQ uses the Viterbi algorithm to select the path which has the minimum mean squared error (MMSE) to the input sequence. In this case we consider a rate-1/2 convolutional encoder shown in the left side of Fig. 6.3, the corresponding branch labeling in trellis is shown in the right side. However, with the codebook $[-1 - 1 - 11]$ and the mapping principle given in (6.5), 1 might be mapped into -1, which is detrimental to the system performance.

In order to solve the problem caused by erroneous mapping in TCQ when source-to-channel SNR is high, we find two solutions which will be explained later. When the source-to-relay channel SNR is high, the soft information approximates either 1 or -1. In such a case, we can consider to use the DF scheme. Hence we set a threshold based on the instantaneous bit error rate (BER) at the relay, if the BER is larger than the threshold, the EF with TCQ scheme is used to deliver the soft information. Otherwise, DF scheme is used to forward the information, in this case the relay makes hard decisions on the symbol x_{S_1} and x_{S_2}, and then calculates the network coded symbol based on the hard decisions $\hat{x}_{S_1}$ and $\hat{x}_{S_2}$, i.e., $\hat{x}_R = \hat{x}_{S_1}\hat{x}_{S_2}$ and forwards it to the destination. The reason for selecting the threshold will be discussed in Section 6.4.

On the other hand, when the source-to-relay channel SNR is high, as the detrimental performance of TCQ results from the unreasonable selection of the codebook, we may refine the codebook to reduce the mapping errors. Motivated by (6.5), we infer that instead of selecting the path which has the MMSE of the input sequence of length m, we can minimize the following equations

$$d(\mathbf{x}, \hat{\mathbf{x}}) = \sum_{i=1}^{m} d_i(x_i, \hat{x}_i),$$

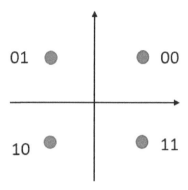

FIGURE 6.4: QPSK constellation.

where
$$d_i(x_i, \hat{x}_i) = \sqrt{(x_i - \hat{x}_i)^2}. \tag{6.7}$$

(6.7) requires that for each state in the trellis, the branches entering or leaving the state should represent all possible codewords. This can be simply achieved by repeating codewords in the codebook. With a rate-1/2 convolutional encoder, the TCQ encoding rate is 1 b/sample, and the size of the codebook is $2^{1+1} = 4$. For the received signals under high SNR region, in order to guarantee the correct mapping for each calculated soft information, codewords need to be repeated, and here only two different codewords can exist in the codebook of length 4, namely 1 and −1. Then the codebook consists of 4 codewords, i.e., −1, −1, 1, and 1. There are several ways to sort the 4 codewords in a codebook which can guarantee accurate quantization, but TCM symbols mapping should be consistent with the order of the codebook. For instance, if Quadrature phase-shift keying (QPSK) with natural mapping shown in Fig. 6.4 is used as the modulation scheme, and branch labeling is depicted on the right side of Fig. 6.3, one possible codebook can be $[-1 1 1 - 1]$. As it can be seen from Fig. 6.4, when −1 is mapped to both 00 and 11 symbols, and 1 is mapped to both 01 and 10 symbols, the overall QPSK constellation can be regarded as a BPSK constellation which expands the distance between different codewords.

6.4 PERFORMANCE ANALYSIS

We first study the threshold for the selection of the forwarding schemes. Then we compare the performance between the two schemes, namely the EF with TCQ and DF scheme, and the EF with TCQ and EF with TCQ codewords repeating scheme.

6.4.1 SET A THRESHOLD

As the distribution of the soft information at the relay is varied according to the source-to-relay channel SNR, we force the relay to evaluate the quality of the received signals by checking whether the received signals satisfy the pre-determined requirement. As we described previously, we set a threshold on the BER of the received data blocks, when the instantaneous BER is larger than the threshold, we consider to use DF scheme, otherwise, we use the EF and TCQ scheme to deliver the soft information.

Design of Soft Network Coding for Two-Way Relay Channels   109

Instantaneous BER for each received block at the relay is computed from (6.8) as

$$P_{e-SR} = \frac{Q\left(\sqrt{E_s h^2_{S_1 R}/\sigma^2_R}\right) + Q\left(\sqrt{E_s h^2_{S_2 R}/\sigma^2_R}\right)}{2}, \quad (6.8)$$

where $Q(\cdot)$ is a Q-function [13, (4.1)], and the content inside the Q-function denotes the SNR of the source-to-relay channel.

For each received m-symbol block, if one symbol is decoded erroneously, the BER is equal to $1/m$. If the instantaneous BER is smaller than $1/m$, the source-to-relay transmission is deemed to be highly reliable and the soft information concentrated on 1 and -1, thus we use DF scheme or EF with TCQ codewords repeating scheme. Otherwise, the EF with conventional TCQ scheme is utilized. Namely,

$$\text{Relay Strategy} = \begin{cases} \text{DF or EF with TCQ codewords repeat,} & \text{if } P_{e-SR} < \frac{1}{m} \\ \text{EF with TCQ,} & \text{if } P_{e-SR} \geq \frac{1}{m} \end{cases}. \quad (6.9)$$

Setting certain BER as a threshold is equivalent to setting the corresponding source-to-channel SNR as a threshold according to (6.8). The threshold only exists for the DF scheme in high SNR region. If the EF with TCQ codewords repeating scheme is used, the threshold no longer exists. This is because the codebook is already pre-determined at both the relay and the receivers based on different source-to-channel SNRs. However, if the DF scheme is used, a threshold is needed for the relay to decide which scheme to choose.

6.4.2 PERFORMANCE ANALYSIS ON THE TWO SCHEMES

In low source-to-relay channel SNR region, in which the instantaneous BER is greater than the threshold, the received signal at the receivers is given by (6.4). It is then multiplied with x_{S_i} to cancel x_{S_i}. As for the EF with TCQ scheme, we follow the method for the EF protocol in [4], it is defined that $\hat{x}_{S_i} = \psi_i(x_{S_i} + e_{S_i})$, where e_{S_i} denotes the uncorrelated soft noise, and ψ_i represents the scalar factors which makes soft noise e_{S_i} uncorrelated to the information symbol x_{S_i}, namely, $\mathrm{E}[e_{S_i} x_{S_i}] = 0$, and $\psi_i = \frac{\mathrm{E}[x_{S_i} \hat{x}_{S_i}]}{\mathrm{E}[x^2_{S_i}]}$. The canceling process can be expressed as

$$x_{S_i} r_{RS_i} = \beta \hat{x}_{S_1} \hat{x}_{S_2} x_{S_i} + n_{equiv\_RS_i} x_{S_i}$$
$$= \beta \psi_1 \psi_2 x_{S_i} + \beta \psi_1 \psi_2 \left(e_{S_j} + x_{S_i} x_{S_j} e_{S_i} + x_{S_i} x_{S_j} e_{S_j}\right) + n_{equiv\_RS_i} x_{S_i}, \quad (6.10)$$

where $\beta = \sqrt{E_R / \mathrm{E}\left[\left|\tanh\left(\frac{LLR_{x_{S_1},R}}{2}\right)\tanh\left(\frac{LLR_{x_{S_2},R}}{2}\right)\right|^2\right]}$.

We define the noise item as $N_{S_i} \triangleq \beta \psi_1 \psi_2 \left(e_{S_j} + x_{S_i} x_{S_j} e_{S_i} + x_{S_i} x_{S_j} e_{S_j}\right) + n_{equiv\_RS_i} x_{S_i}$, and we regard it as an approximately Gaussian distributed random variable with mean value $m_{N_{S_i}}$ and variance $\sigma^2_{N_{S_i}}$. According to [14], the mean value of the output of a minimum mean-square error quantizer is equal to the mean value of the input. Due to the linearity of the expected value, the mean value of the quantization noise is 0. Besides, since $\mathrm{E}[e_{S_i} x_{S_i}] = 0$, we can get $m_{N_{S_i}} = 0$ and $\sigma^2_{N_{S_i}} = \beta^2 \psi^2_1 \psi^2_2 \left(\sigma^2_{e_{S_1}} + \sigma^2_{e_{S_2}} + \sigma^2_{e_{S_1}} \sigma^2_{e_{S_2}}\right) + \sigma^2_{equiv\_RS_i}$, where $\sigma^2_{e_{S_i}}$ denotes the variance of the soft noise e_{S_i}, and $\sigma^2_{equiv\_RS_i}$ represents the variance of the quantization noise.

In the high source-to-relay channel SNR region, if the DF scheme is used, the hard decision value, which is calculated by $\hat{x}_R = \hat{x}_{S_1}\hat{x}_{S_2}$, is forwarded at the relay. At S_i, the MMSE estimation of x_{S_i} is multiplied with x_{S_i} in order to cancel x_{S_i}, which can be expressed as $x_{S_i} r_{RS_i} = h_{RS_i}\sqrt{E_R}\hat{x}_R x_{S_i} + n_{RS_i} x_{S_i}$. If the EF with TCQ codewords repeating scheme is used, it follows the same steps as the EF with TCQ scheme.

The main difference between the two schemes is that, at the receiver, the received signal of the DF scheme is the superposition of the useful signal $\hat{x}_R$ multiplied with the channel coefficient h_{RS_i} and the receiver noise n_{RS_i}. While for the EF with TCQ codewords repeating scheme, except for occasional error events resulting from deep fading of the relay-to-source channel, the received signal can be correctly mapped to the codewords 1 or −1 without the superposition of the receiver noise or the multiplication of the channel coefficient, but instead the soft noise e_{S_i} and the quantization noise $n_{equiv\_RS_i}$ are induced. The other difference is in that for the TCQ combined with the DF scheme, we need to switch between two different schemes based on certain threshold, while for the other scheme, no need to set the threshold.

6.5 SIMULATION RESULTS

In the simulation, we consider Rayleigh fading channels, all the channel coefficients are modeled by complex Gaussian random variables. Furthermore, the channel coefficients between the sources and the relay, viz. $h_{S_1 R}$, $h_{S_2 R}$, h_{RS_1}, h_{RS_2}, have zero mean value and unit variance. The channel coefficients between sources, i.e, $h_{S_2 S_1}$, $h_{S_2 S_1}$ have zero mean value and the variance of 0.36. We further assume that each data block contains 1000 binary symbols. Moveover, we assume that the transmission power and the receiver noise variance at both the relay and sources are equal. At the relay, if the instantaneous BER of the received signal is less than the threshold $1/1000$, either the DF scheme or the EF with the TCQ codewords repeating scheme is applied. Otherwise, conventional EF with TCQ scheme is used. Finally, the system BER performance is used to measure the quality of each scheme. By "system BER performance", we mean that the average value of the two sources' BERs.

In the simulations, we take the AF, DF and EF schemes as benchmarks. Fig. 6.5 shows the scenario in which the DF scheme is combined with the EF plus TCQ scheme when instantaneous BER of the received signal is less than $1/1000$. We can see that the EF with TCQ combined with DF scheme achieves full diversity gain, i.e., 2 for TWRC, and outperforms both of the conventional DF and AF schemes.

Fig. 6.6 shows the BER performance for the scenario in which the EF with TCQ codewords repeating scheme is combined with the EF plus conventional TCQ scheme. We are using QPSK with natural mapping modulation scheme in TCM. As discussed in Section 6.3, a codebook of $[-111-1]$ for TCQ is consistent with natural QPSK constellation, thus we chose this codebook in our simulation. As can be seen from Fig. 6.6, the conventional EF with TCQ combined with the EF with codewords repeating scheme outperforms both the AF and the conventional DF schemes. As the SNR increases, the probability that the instantaneous BER of the received signals falls below the threshold increases, thus the times of the system operating at the EF with TCQ codewords repeating scheme will grow. Since trellis code is used in the quantization, the quantized and modulated signals can be viewed as coded signals, at the receiver, the received signal can be detected and mapped backed to the codewords −1 or 1, this gives better performance than the EF with TCQ combined with the DF scheme. It can also be seen that the performance of this scheme almost overlaps with the conventional EF scheme in which soft information is directly delivered to the destination.

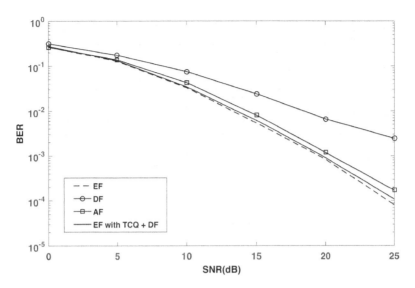

FIGURE 6.5: BER performance for fading channels.

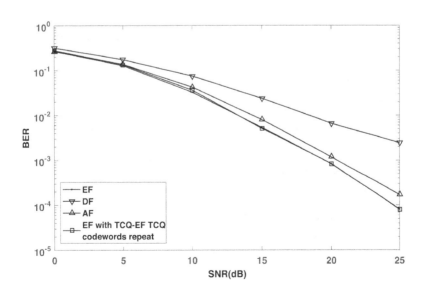

FIGURE 6.6: BER performance for fading channels.

6.6 CONCLUSION

This chapter presents one possible means to transmit the network coded soft information for the EF protocol in TWRC at rate $R = 1$ b/sample. We take advantage of joint trellis coded quantization/trellis coded modulation systems to forward the soft information when the source-to-channel SNR is low, and when the source-to-channel SNR is high enough to meet the pre-determined requirement, we propose to use either the DF scheme or the EF with TCQ codewords repeating scheme to implement the information delivery at the relay. We find both schemes outperform the AF scheme and conventional DF schemes over two way relay slow fading channels.

1. J. N. Laneman, D. N. C. Tse, and G. W. Wornell, "Cooperative diversity in wireless networks: Efficient protocols and outage behavior," *Information Theory, IEEE Transactions on*, vol. 50, no. 12, pp. 3062–3080, 2004.
2. C. Deqiang and J. N. Laneman, "Modulation and demodulation for cooperative diversity in wireless systems," *Wireless Communications, IEEE Transactions on*, vol. 5, no. 7, pp. 1785–1794, 2006.
3. I. Abou-Faycal and M. Medard, "Optimal uncoded regeneration for binary antipodal signaling," in *Communications, 2004 IEEE International Conference on*, vol. 2, 2004, pp. 742–746 Vol.2.
4. G. Krishna Srikanth and J. Syed Ali, "Optimal relay functionality for snr maximization in memoryless relay networks," *Selected Areas in Communications, IEEE Journal on*, vol. 25, no. 2, pp. 390–401, 2007.
5. M. A. Karim, Y. Tao, Y. Jinhong, C. Zhuo, and I. Land, "A novel soft forwarding technique for memoryless relay channels based on symbol-wise mutual information," *Communications Letters, IEEE*, vol. 14, no. 10, pp. 927–929, 2010.
6. C. Tao, T. Ho, and J. Kliewer, "Memoryless relay strategies for two-way relay channels," *Communications, IEEE Transactions on*, vol. 57, no. 10, pp. 3132–3143, 2009.
7. P. Jung Min, K. Seong-Lyun, and C. Jinho, "Hierarchically modulated network coding for asymmetric two-way relay systems," *Vehicular Technology, IEEE Transactions on*, vol. 59, no. 5, pp. 2179–2184, 2010.
8. C. Hausl and J. Hagenauer, "Iterative network and channel decoding for the two-way relay channel," in *Communications, 2006. ICC '06. IEEE International Conference on*, vol. 4, 2006, pp. 1568–1573.
9. S. Zhang, Y. Zhu, and S. C. Liew, "Soft network coding in wireless two-way relay channels," *J. Communication and Networks, Special Issues on Network Coding*, vol. 10, no. 4, 2008.
10. M. W. Marcellin and T. R. Fischer, "Trellis coded quantization of memoryless and gauss-markov sources," *IEEE Transactions on Communications*, vol. 38, no. 1, pp. 82–93, 1990.
11. L. Zihuai and T. Aulin, "Joint source-channel coding using combined tcq/cpm: iterative decoding," *Communications, IEEE Transactions on*, vol. 53, no. 12, pp. 1991–1995, 2005.
12. T. R. Fischer and M. W. Marcellin, "Joint trellis coded quantization modulation," *IEEE Transactions on Communication*, vol. 39, no. 2, pp. 172–176, 1991.
13. M. Dan, "Digital communication over fading channels, second edition," p. 160, 2005.
14. J. Bucklew and J. Gallagher, N., "A note on optimal quantization (corresp.)," *Information Theory, IEEE Transactions on*, vol. 25, no. 3, pp. 365–366, 1979.

7 Linear Neighbor Network Coding

7.1 INTRODUCTION

Wireless communications between devices can be unreliable owing to a number of issues, such as channel fading, interference or mobility of devices. Therefore, transmission reliability is a major challenge in wireless communications [1–4]. This work examines the reliability of broadcasting in wireless networks using network coding. Conventionally, in a lossy wireless network without network coding, the successful reception of a packet relies on multiple retransmissions of the same information.Therefore, in literature, reliability is usually measured in terms of the number of retransmissions. Numerous research has been conducted to reduce the number of retransmissions, such as Automatic Repeat reQuest (ARQ) and Forward Error Correction (FEC). Most recently, network coding is employed to further reduce the number of retransmissions.

For reliable broadcast, network coding has frequently been used as an error control technique. It increases network reliability by reducing retransmissions [1–4]. In [1,2], network coding aided ARQ is investigated in access point (AP) based networks. There, network coding is used to broadcast a specific set of unsuccessfully received packets to various receivers. Users in citenc ARQ can listen to all packets. As a result, using the knowledge of overheard packets, intended users can decode the network coded packet. In terms of service time and goodput, citexor rescue considers the fairness of all users. The chapter also demonstrates the effectiveness of the network coding aided retransmission scheme in a real-world setting.

Network coding is used in [3,4] for networks with tree topology and an equal number of children in each multicast tree. The number of retransmissions expected by the source node under various error control protocols is calculated. It is hypothesized that network coding achieves a logarithmic reliability gain with respect to the number of receivers in a multicast group when compared to a simple ARQ scheme based on numerical comparison. The latter work then proves this hypothesis.

Unlike previous research, this study considers an all-to-all broadcasting model. Instead of the expected number of retransmissions, this work gives the exact probability that after each retransmission, every node in the network receives packets successfully from every other node. Furthermore, in the network, feedback is not required.

Nistor et al. investigate the delay probability distribution of message broadcast in wireless networks using random linear network coding in [5]. At individual delay, the probability of successful decoding is similar to the reliability discussed in this chapter. They investigate a network in which a single source sends multiple packets over erasure channels to multiple receivers. They analyse a network with only two receivers using a Markov chain model, and a network with three receivers using a brute-force method. This work, on the other hand, considers any number of receivers.

The first section of this chapter looks at a neighbor network coding scheme for lossy wireless networks. Then a Markov chain model is created, which is used to obtain exact analytical reliability results, where reliability is defined as the probability that every node in the network receives packets from every other node. It is demonstrated that the proposed neighbor network coding scheme improves network reliability. To better reveal the impact of network parameters on reliability, closed-form upper and lower bounds on reliability using the proposed coding scheme are also provided. Simulations are used to validate the analysis.

The remainder of the chapter is laid out as follows. The system model is introduced in section refsection network model. In Section 7.3, the theoretical analysis of reliability is presented, followed by the closed-form bounds on reliability in Section 7.4. The simulation and numerical results are presented in Section 7.5. The chapter is concluded with Section 7.6, which proposes future work.

7.2 SYSTEM MODEL

The kth node in a network with n nodes is denoted by N_k. Each node serves as a source node, broadcasting a packet to all other nodes in the network. N_k broadcasts the original packet, which is denoted by x_k. Furthermore, it is assumed that time is slotted, and that only one source node (say, N_j) broadcasts a single packet during each time slot.

The packet broadcast from a source node may not be able to reach every other node in one time slot due to the lossy nature of wireless communications. Let p_{ji} be the probability that a packet broadcast from N_j reaches N_i successfully in one time slot. Note that p_{ji} includes the probabilities that the packet broadcast from N_j reaches N_i via either a single hop path (i.e., a link) if exists, a multi-hop path or multiple paths.

The *probabilistic connectivity matrix*, which is used to obtain p_{ji} for all pairs of nodes in a network, is described in [7]. Because the focus of this research is on the impact of network coding on reliability, p_{ji} is assumed to be known. Further, we assume that p_{ji} does not change over time and that the events associated with p_{ji}, $1 \leq i, j \leq n, i \neq j$ are independent.

If a source node's packet broadcast does not reach all nodes in a single time slot, the source node must broadcast multiple times. Assume that all nodes in the network transmit round robin and that a successful transmission is not acknowledged, which is a typical broadcast scenario. A *round* is a set of time slots in which each source node broadcasts exactly once. As a result, at round R (time slot $t = nR$), the network's *reliability* is defined as the probability that every node in the network has a copy of the packets broadcast by all other nodes at round R.

It's worth mentioning that a source node can only re-broadcast its original packet to boost dependability without network coding. The source node can broadcast a combination of its original packet and received packets using network coding. *neighbor network coding scheme* is the network coding scheme used in this study. Each node (say, N_j) chooses another node (say, N_h), namely the *coding neighbor*, to perform the XOR coding, where $j \neq h$. A lossy wireless network using neighbor network coding is shown in Fig. 7.1. Each node has a buffer that stores all of the received packets. After each packet is received, decoding is performed at each node. N_j broadcasts $x_j \oplus x_h$ if it has x_h in its buffer; otherwise, it broadcasts x_j. As a result, the packet broadcast by node N_j at time t is determined by the packets received by N_j from other source nodes prior to time t. As shown in the next section, this puts the theoretical analysis to the test.

7.3 THEORETICAL ANALYSIS

In this section, we study the reliability of the network by examining the packets received by a node (say N_i) from an arbitrary source node (say N_j).

Suppose that the coding neighbor of N_j is N_h. Then, N_j may broadcast either x_j or $x_j \oplus x_h$, depending on the packets that N_j has. It follows that the state of the node N_i at a time slot, viz. the packets received by N_i, depends only on the states of N_i and N_j at the previous time

Linear Neighbor Network Coding

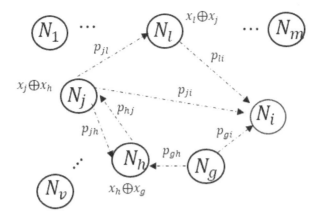

FIGURE 7.1: Illustration of a lossy wireless network with n nodes applying neighbor network coding. Note that some paths are not shown in the figure.

slot *and* the packet receptions at this time slot, where the probability distribution of the latter one is time-invariant due to the fact that p_{ji} is time-invariant. Therefore, the transmission can be modeled by a Markov chain.

7.3.1 CONSTRUCTION OF THE STATES

Let the *state* of a node be the categories of its received packets. More specifically, the state is expressed by a vector consisting of n elements, i.e. $[\xi_1, ..., \xi_n]$, where ξ_k indicates the categories of packets received from node N_k. There are four possible categories can be received from each source node. These are: no packet, original packet, XORed packet, and both original and XORed packets, represented by $\xi_k = 0, 1, 2,$ and 3 respectively. Note that $\xi_i = 1$ in the states of N_i, because N_i always has its own packet. The total number of states L for each node is equal to 4^{n-1}.

After receiving the packet from its designated coding neighbor, a source node begins broadcasting the XORed packet. As a result, receiving the original packet (x_j) from a source node (N_j) after receiving the XORed packet $x_j \oplus x_h$ is impossible. As a result, there are some absorbing states that cannot be exited once entered. For all $k \in \{1, 2, ..., n\} \setminus i$, the absorbing states of node N_i have the property $\xi_k = 2$ or 3.

As an example, consider a network with three nodes. Assume that the coding neighbors for $N_1, N_2,$ and N_3, respectively, are $N_3, N_1,$ and N_2. Each node has a total of $L = 16$ states. Table 7.1 shows the states of N_1 and their corresponding packets in the buffer as an example.

7.3.2 TRANSITION MATRICES

Consider the fact that N_j broadcasts during the time slot t. The transition of the states of a receiving node, say N_i, is then investigated. a represents the state of N_i at time t, and b represents the state of N_i at time $t+1$.

The transition matrix governing the transitions of the states of N_i when N_j broadcasts is denoted by $Q_{ji}(t)$. It's worth noting that $Q_{ji}(t)$ is dependant on the packet N_j

TABLE 7.1

The states of N_1 and corresponding packets for a network of three nodes, with N_3, N_1, and N_2 being the coding neighbours for N_1, N_2, and N_3, respectively. The 6th state, for example, is $\{111\}$, which indicates that N_1 has packets x_1, x_2 and x_3.

Index	1	2	3	4
State	$\{100\}$	$\{101\}$	$\{102\}$	$\{103\}$
Packets	x_1	x_1, x_3	$x_1, x_3 \oplus x_2$	$x_1, x_3, x_3 \oplus x_2$
Index	5	6	7	8
State	$\{110\}$	$\{111\}$	$\{112\}$	$\{113\}$
Packets	x_1, x_2	x_1, x_2 x_3	x_1, x_2 $x_3 \oplus x_2$	x_1, x_2 $x_3, x_3 \oplus x_2$
Index	9	10	11	12
State	$\{120\}$	$\{121\}$	$\{122\}$	$\{123\}$
Packets	$x_1, x_2 \oplus x_1$	$x_1, x_2 \oplus x_1$ x_3	$x_1, x_2 \oplus x_1$ $x_3 \oplus x_2$	$x_1, x_2 \oplus x_1$ $x_3, x_3 \oplus x_2$
Index	13	14	15	16
State	$\{130\}$	$\{131\}$	$\{132\}$	$\{133\}$
Packets	x_1 x_2 $x_2 \oplus x_1$	x_1 $x_2, x_2 \oplus x_1$ x_3	x_1 $x_2, x_2 \oplus x_1$ $x_3 \oplus x_2$	x_1 $x_2, x_2 \oplus x_1$ $x_3, x_3 \oplus x_2$

broadcasts, which could be the original or XORed packet. As a result, denote $M_{ji}^{\mu_1}$ and $M_{ji}^{\mu_2}$ as conditional transition matrices representing the transition matrices of the state of N_i conditioned on the event that N_j broadcasts its original and XORed packets, respectively. Let $M_{ji}^{\mu_1}$ and $M_{ji}^{\mu_2}$ be $L \times L$, for example:

$$M_{ji}^{\mu_1} = \begin{bmatrix} P_M^{\mu_1}(1|1) & \cdots & P_M^{\mu_1}(L|1) \\ \vdots & \ddots & \vdots \\ P_M^{\mu_1}(1|L) & \cdots & P_M^{\mu_1}(L|L) \end{bmatrix}, \quad (7.1)$$

where each element of $M_{ji}^{\mu_1}$ and $M_{ji}^{\mu_2}$, denoted by $P_M^{\mu_1}(b|a)$ and $P_M^{\mu_2}(b|a)$ respectively, is the probability that the state of N_i changes from a to b during time slot t conditioned on the event that N_j broadcasts its original packet and the XORed packet respectively.

Then, $Q_{ji}(t)$ can be computed as follows according to total probability theory:

$$Q_{ji}(t) = M_{ji}^{\mu_1} \mu_1(t) + M_{ji}^{\mu_2} \mu_2(t), \quad (7.2)$$

where $\mu_1(t)$ (resp. $\mu_2(t)$) is the probability that N_j transmit its original packet (resp. the XORed packet) at time slot t.

In the next section, the probabilities $\mu_1(t)$ and $\mu_2(t)$ will be discussed. The time-invariant conditional transition matrices can be constructed using the following rules.

Linear Neighbor Network Coding

Each entry of $M_{ji}^{\mu_1}$, say $P_M^{\mu_1}(b|a)$, can be constructed by comparing states a and b, according to Algorithm 7.1. In the algorithm, $a\{k\}$ denotes the kth element of state a and we say $a = b$ if $a\{k\} = b\{k\}$ for all $k \in \{1, 2, \ldots, n\}$. Similarly, each entry of $M_{ji}^{\mu_2}$, say $P_M^{\mu_2}(b|a)$, can be constructed by comparing the states a and b, according to Algorithm 7.2.

Algorithm 7.1 Construct $M_{ji}^{\mu_1}$

---

  for each $P_M^{\mu_1}(b|a)$ in $M_{ji}^{\mu_1}$ **do**
    if $a = b$ and $a\{j\} = b\{j\} = 0$ **then** N_i does not receive the packet from N_j, which happens with probability $P_M^{\mu_1}(b|a) = 1 - p_{ji}$;
    else if $a\{j\} = 0$, $b\{j\} = 1$, while $a\{k\} = b\{k\}$ for all $k \in \{1, 2, \ldots, n\} \setminus j$ **then** N_i receives the packet from N_j, which happens with probability $P_M^{\mu_1}(b|a) = p_{ji}$;
    else if $a = b$ and $a\{j\} = b\{j\} \neq 0$ **then** the state transition does not depend on whether or not N_i receives the packet from N_j, hence $P_M^{\mu_1}(b|a) = 1$;
    else let $P_M^{\mu_1}(b|a) = 0$.
    end if
  end for

Algorithm 7.2 Construct $M_{ji}^{\mu_2}$

---

  for each $P_M^{\mu_2}(b|a)$ in $M_{ji}^{\mu_2}$ **do**
    if $a = b$ and $a\{j\} = b\{j\} = 0$ or 1 **then** N_i does not receive the packet from N_j, which happens with probability $P_M^{\mu_2}(b|a) = 1 - p_{ji}$;
    else if $a\{j\} = 0$ and $b\{j\} = 2$ **or** $a\{j\} = 1$ and $b\{j\} = 3$, while $a\{k\} = b\{k\}$ for all $k \in \{1, 2, \ldots, n\} \setminus j$ **then** N_i receives the packet from N_j, which happens with probability $P_M^{\mu_2}(b|a) = p_{ji}$;
    else if $a = b$ and $a\{j\} = 2$ or 3 **then** the state transition does not depend on whether or not N_i receives the packet from N_j, hence $P_M^{\mu_2}(b|a) = 1$;
    else let $P_M^{\mu_2}(b|a) = 0$.
    end if
  end for

7.3.3 THE STATE VECTORS

$S_i(t)$ denotes the state vector of node N_i at time slot t. A state vector is a $1 \times L$ row vector in which the lth entry represents the likelihood that N_i is in the lth state at time slot t. If N_j broadcasts at time slot t, the state vector of N_i at time slot $t+1$ can be calculated using Eq. (7.2):

$$\begin{aligned} S_i(t+1) &= S_i(t) Q_{ji}(t) \\ &= S_i(t) \left(M_{ji}^{\mu_1} \mu_1(t) + M_{ji}^{\mu_2} \mu_2(t) \right). \end{aligned} \quad (7.3)$$

The next step is to get $\mu_1(t)$ and $\mu_2(t)$. B_j is a $L \times 1$ indicator vector, with the lth entry set to one if N_j broadcasts the XORed packet in the lth state, and zero otherwise. Let A_j be a $L \times 1$ indicator vector, with the lth entry set to one if N_j broadcasts its original packet in the lth state and zero otherwise.

As a result, a recursive formula containing the state vectors of N_i and N_j at time slot t can be used to generate the state vector of N_i at time slot $t + 1$:

$$S_i(t+1) = S_i(t)\left(S_j(t) \times A_j \times M_{ji}^{\mu_1} + S_j(t) \times B_j \times M_{ji}^{\mu_2}\right). \quad (7.4)$$

The initial state of N_i contains packet x_i only. The initial state is then assigned with probability one in the initial state vector $S_i(0)$, while all other states are assigned with probability zero. For example, if the states of N_1 are arranged as shown in Table 7.1, the initial state is [100]. As a result, $S_i(0)$ is a 1×16 vector with one entry and zero entries.

7.3.4 RELIABILITY

Denote by $\psi_i(t)$ the probability that node N_i has packets from every other node at time slot t. Then, it can be calculated by:

$$\psi_i(t) = \sum_{x \in \chi} S_i^x(t), \quad (7.5)$$

where $S_i^x(t)$ is the xth entry of $S_i(t)$, The indexes of states in which N_i has the packets from every other node are included in χ. As an example, for $i = 1$, we have $\chi = \{4, 6, 7, 8, 10, 11, 12, 14, 15, 16\}$, as shown in Table 7.1.

Finally, the network's reliability at time slot t, i.e., the probability that each node receives packets from each other, can be expressed as follows:

$$\psi(t) = \prod_{i \in \{1,2,\ldots,n\}} \psi_i(t). \quad (7.6)$$

7.3.5 NETWORKS WITHOUT NETWORK CODING

We also look at the reliability of networks that don't use network coding as a comparison. The reliability can be calculated in the same way as the previous example.

The state of a node can also be represented by a $1 \times n$ row vector, where the kth element takes value either 0 or 1, indicating whether the node has received or has not received xk, similar to the construction of states in sub-section 7.3.1. Eq. (7.3) can then be used to generate the state vector of a node (say Ni).

Furthermore, because each node only transmits its original packet, the transition matrix Q_{ji} is time-invariant. Define $Q \triangleq Q_{11}Q_{21}\cdots Q_{n1}$. The state vector of N_i at time slot $t = nR$ can then be calculated using the following formula:

$$S_i(t) = S_i(0)Q^R, \quad (7.7)$$

where $S_i(0)$ is the initial state vector of N_i at time 0, as introduced in sub-section 7.3.3.

The network's reliability can then be calculated using Eqs. (7.5) and (7.6), where χ is the index of the state in which all elements are 1, which corresponds to a node receiving packets from all other nodes in the network.

7.4 BOUNDS ON THE RELIABILITY

Although the theoretical results presented in the preceding section are exact, the computation can be difficult. In this section, we present closed-form results of upper and lower bounds on network reliability to shed more light on the impact of fundamental network parameters, such as connectivity between nodes p_{ij} and coding neighbor selection, on network reliability.

The analysis begins with a single packet x_j being received at a single node N_i. Assume that node N_j chooses N_h as its coding neighbor, and that node N_d chooses N_j as its coding neighbor, as shown in Fig. 7.1. The packet x_j can then reach N_i through one of two processes. The first is via a path N_j to N_i, via the reception of packets x_j or $x_j \oplus x_h$; the second is via a path N_j to N_d, followed by a path N_d to N_i, via the reception of packet $x_d \oplus x_j$.

Denote $F_{ji}(R)$ as the probability that N_i receives and decodes x_j by round R, and $f_{ji}(R)$ as the probability that N_i receives and decodes x_j at round R.

7.4.1 THE UPPER BOUND

Theorem 1. *Suppose that N_d selects N_j as coding neighbor. The probability that node N_i receives x_j in R rounds satisfies:*

$$F_{ji}(R) \leq (1-p_{ji})^R \sum_{\alpha=1}^{R} \left(1 - (1-p_{di})^{R-\alpha}\right)(1-p_{jd})^{\alpha-1} p_{jd}$$
$$+ \left(1 - (1-p_{ji})^R\right) \triangleq U_{ji}(R). \quad (7.8)$$

Proof: We consider that N_i can decode x_j upon receiving any packet from N_j, regardless of whether the packet is the original x_j or the XORed packet, to get an upper bound on the probability $F_{ji}(R)$. Note that even if N_i receives a packet broadcast from N_j, it may still be unable to decode x_j. This is because N_j may broadcast $N_j \oplus N_h$, and whether or not N_i has x_h in its buffer determines whether or not x_j is successfully decoded.

The event that a packet containing x_j (either x_j or an XORed packet containing x_j) reaches N_i by round R via the first (resp. second) process is denoted by Ξ_R (resp. Γ_R). As a result, it is clear that $F_{ji}(R) \leq \Pr(\Xi_R \cup \Gamma_R) = \Pr(\Xi_R) + (1 - \Pr(\Xi_R))\Pr(\Gamma_R)$.

Further, t $\Pr(\Xi_R) = 1 - (1-p_{ji})^R$ and

$$\Pr(\Gamma_R) = \sum_{\alpha=1}^{R} \left(1 - (1-p_{di})^{R-\alpha}\right) f_{jd}(\alpha), \quad (7.9)$$

where α is the round in which the packet broadcast by N_j first reaches N_d. It is clear that $f_{jd}(\alpha)$ follows a geometric distribution with p_{jd} as the success probability.
Then:

$$\Pr(\Gamma_R) = \sum_{\alpha=1}^{R} \left(1 - (1-p_{di})^{R-\alpha}\right)(1-p_{jd})^{\alpha-1} p_{jd}. \quad (7.10)$$

Then, Eq. (7.8) can be readily obtained. ∎

Similarly to Eq. (7.6), the Rth round network reliability upper bound, is given by:

$$U(R) = \prod_{i,j \in \{1,2,...,n\}} U_{ji}(R), \quad (7.11)$$

where $U_{ji}(R)$ is given by Theorem 1.

7.4.2 THE LOWER BOUND

Theorem 2. *Suppose that the coding neighbors of N_j, N_d, and N_h are N_h, N_j, and N_g respectively. The probability that node N_i has packet x_j at the Rth rounds satisfies:*

$$F_{ji}(R) \geq \sum_{\beta=1}^{R} f_{hj}^{L}(\beta) \left(\Pr(\Omega_R|\beta) + \Pr(\Psi_R|\beta) - \Pr(\Omega_R|\beta)\Pr(\Psi_R|\beta) \right), \quad (7.12)$$

where $\Pr(\Omega_R|\beta)$, $\Pr(\Psi_R|\beta)$ and $f_{hj}^{L}(\beta)$ are given by Eq. (7.14), Eq. (7.15) and Eq. (7.17) respectively.

Proof: We investigate the two processes described at the beginning of this section separately, as we did with the proof of Theorem 1. Define α as the round in which N_d receives x_j for the first time and begins broadcasting $x_d \oplus x_j$. Denote β as the round in which N_j receives x_h from N_h for the first time and begins broadcasting $x_j \oplus x_h$. We only consider cases where N_h broadcasts its original packet in the first β rounds to get a lower bound on network reliability, and we ignore the probability that N_h broadcasts coded packets.

In the first process, it is self-evident that a N_j XORed packet, i.e., $x_j \oplus x_h$, can be decoded by N_i if N_i has packet x_h. The event that N_i receives the packet x_j via the first process by round R is denoted by Ω_R^A. Denote the event that N_i receives the packet $x_j \oplus x_h$ via the first process by round R but N_i only stores the packets received from N_h in the first β rounds by Ω_R^B.

Let $\Pr(\Omega_R^A|\beta)$ be the event probability that Ω_R^A occurs if β is the round in which node N_j receives x_h for the first time. It is simple to understand, $\Pr(\Omega_R^A|\beta) = 1 - (1-p_{ji})^\beta$. Similarly,

$$\Pr(\Omega_R^B|\beta) = \left(1-(1-p_{ji})^{R-\beta}\right)\left(1-(1-p_{hi})^\beta + (1-p_{hi})^\beta\left(1-(1-p_{ji})^\beta\right)\right). \quad (7.13)$$

The first multiplication term is the probability that N_i will receive $x_j \oplus x_h$, and the second multiplication term is the probability that N_i will receive either x_j or x_h, allowing the XORed packet to be decoded.

Further, $\Pr(\Omega_R^A \cap \Omega_R^B) = \Pr(\Omega_R^B|\Omega_R^A)\Pr(\Omega_R^A) = \left(1-(1-p_{ji})^{R-\beta}\right)\left(1-(1-p_{ji})^\beta\right)$ due to the correlated events Ω_R^A and Ω_R^B. Finally, we can have a lower bound on the probability that N_i receives and decodes x_j at round R:

$$\begin{aligned}\Pr(\Omega_R|\beta) &\triangleq \Pr(\Omega_R^A \cup \Omega_R^B|\beta) \quad (7.14)\\ &= \Pr(\Omega_R^A|\beta) + \Pr(\Omega_R^B|\beta) - \Pr(\Omega_R^A \cap \Omega_R^B|\beta).\end{aligned}$$

Both $x_d \oplus x_j$ and x_d are required for N_i to decode x_j in the second process. Ψ_R denotes the event in which N_i receives x_j via the second process but N_i only receives x_d from N_d when N_d broadcasts its original packet. Then, given that N_j receives x_h for the first time at

Linear Neighbor Network Coding

round β, the probability that event Ψ_R occurs is:

$$\Pr(\Psi_R|\beta) = \sum_{\alpha=1}^{\beta}(1-(1-p_{di})^\alpha)\left(1-(1-p_{di})^{R-\alpha}\right)f_{jd}(\alpha) \tag{7.15}$$

$$= \sum_{\alpha=1}^{\beta}(1-(1-p_{di})^\alpha)\left(1-(1-p_{di})^{R-\alpha}\right)\left((1-p_{jd})^{\alpha-1}p_{jd}\right),$$

where the first term in the summation is the probability that N_i will receive x_d; the second term is the probability that N_i will receive the XORed packet $x_d \oplus x_j$; and the third term is the geometric distribution with success probability p_{jd} as introduced in the proof of Theorem 1.

Therefore, the probability that node N_i receives x_j in R rounds satisfies:

$$F_{ji}(R) \geq \sum_{\beta=1}^{R}\phi_{hj}(\beta)\Pr(\Omega_R \cup \Psi_R|\beta) \tag{7.16}$$

$$= \sum_{\beta=1}^{R}\phi_{hj}(\beta)\left(\Pr(\Omega_R|\beta) + \Pr(\Psi_R|\beta) - \Pr(\Omega_R|\beta)\Pr(\Psi_R|\beta)\right),$$

where $\phi_{hj}(\beta)$ is the probability that N_j receives x_h from N_h at round β for the first time, which satisfies:

$$\phi_{hj}(\beta) \geq (1-p_{hj})^{\beta-1}p_{hj}(1-F_{gh}(\beta)) \triangleq f_{hj}^L(\beta), \tag{7.17}$$

where the first term is the likelihood that N_j will not receive x_h by round $\beta-1$, the second term is the likelihood that N_j will receive a packet from N_h at round β, and the third term is the likelihood that N_h will broadcast its original packet at round β. It's worth noting that Theorem 1 can be used to calculate $F_{gh}(\cdot)$.

7.5 RESULTS AND DISCUSSION

Simulations are carried out in this section to validate our theoretical analysis. The advantages of neighbor network coding over non-coded networks are demonstrated, followed by discussions of the coding scheme's relationship to network reliability.

Theoretically, Eq. (7.6) can be used to calculate the reliability of networks with any number of nodes at any round. The probabilistic connectivity matrices representing channel conditions can be arbitrary. In this section, the entries are chosen randomly to generate numerical results. In Fig. 7.2, the theoretical results for networks with 3, 4 and 5 nodes from round $R = 1$ to $R = 10$ are plotted. The coding scheme is that each node selects its index neighbor as coding neighbor, i.e., N_k chooses $N_{k+1 \pmod{n}}$ as coding neighbor. It demonstrates that the theoretical results closely match the simulation results, thereby validating theoretical analysis.

$$Q_5 = \begin{bmatrix} 1 & 0.3 & 0.6 & 0.5 & 0.4; \\ 0.4 & 1 & 0.5 & 0.7 & 0.3; \\ 0.7 & 0.4 & 1 & 0.3 & 0.5; \\ 0.3 & 0.6 & 0.4 & 1 & 0.6; \\ 0.6 & 0.5 & 0.3 & 0.4 & 1 \end{bmatrix}. \tag{7.18}$$

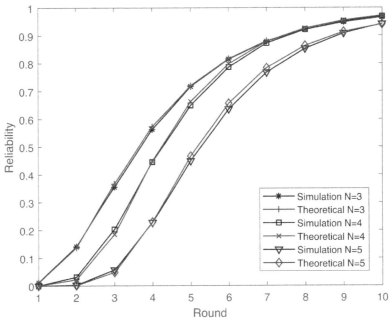

FIGURE 7.2: Simulation and theoretical results of the reliability of networks when $n = 3, 4, 5$, where the probabilistic connectivity matrix is given in eq. 7.18.

In order to examine the reliability gain that neighbor network coding brings, networks with and without coding are plotted in Fig. 7.3. The networks consist of four nodes, with equal probabilities of p_{ij} for every $i, j \in \{1, 2, \ldots, n\}, i \neq j$. Three sets of p_{ij} are chosen, which are $p_{ij} = 0.2$, $p_{ij} = 0.35$ and $p_{ij} = 0.5$. The coding scheme is the same as that in Fig. 7.2. It is obvious that when the number of rounds is small (resp. large), the reliability is low (resp. high). When the number of rounds is moderate, it can be seen that networks applying neighbor network coding outperform non-coded networks and improve the reliability by around 10 percents.

Fig. 7.4 shows the comparison of the probabilities that x_1 is received by the whole network when N_1 chooses different coding neighbors. The connectivity matrices for $n = 3$ and $n = 4$ are given by Q_3 and Q_4 below,

$$Q_3 = \begin{bmatrix} 1 & 0.15 & 0.3 \\ 0.4 & 1 & 0.5 \\ 0.6 & 0.4 & 1 \end{bmatrix}; Q_4 = \begin{bmatrix} 1 & 0.1 & 0.3 & 0.5 \\ 0.2 & 1 & 0.5 & 0.4 \\ 0.4 & 0.3 & 1 & 0.4 \\ 0.3 & 0.2 & 0.1 & 1 \end{bmatrix}. \quad (7.19)$$

The results suggest that the selection of neighbors affects the performance of a single node as well as the whole network. For example, if N_j selects the node to which the connection is the weakest as coding neighbor, the probability of successful reception of x_j by the whole network can be maximized in every round. This conjecture can be applied to optimize the reception of packets from a special node whose information is important for the network.

Linear Neighbor Network Coding

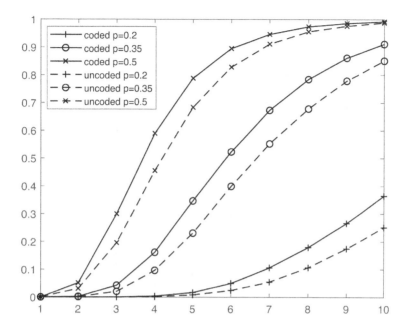

FIGURE 7.3: The reliability gain of neighbor coding network over non-coded network when $n = 4$.

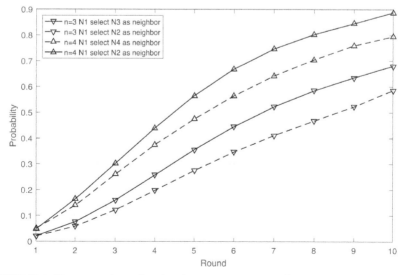

FIGURE 7.4: The reception of x_1 by the network with different neighbor network coding schemes, where the connectivity matrices are given in Eq. 7.19.

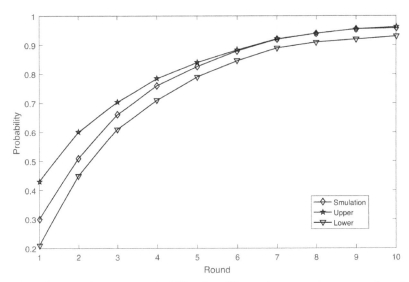

FIGURE 7.5: Bounds on the probability that N_3 receives x_1 where $n = 3$ nodes and the connectivity matrix is given by 7.20.

Theorems 1 and 2 give bounds on the probability that x_1 is received by N_3, which are shown in Fig. 7.5 using the same coding scheme in Table 7.1. The connectivity matrix is given by

$$Q'_3 = \begin{bmatrix} 1 & 0.2 & 0.3 \\ 0.4 & 1 & 0.1 \\ 0.7 & 0.4 & 1 \end{bmatrix}. \tag{7.20}$$

The bounds are valid, as can be seen. The bounds, on the other hand, can be improved and used to characterize the reliability gain as well as to facilitate the proof of the aforementioned conjecture, which is a difficult task that will be left for future work.

7.6 CONCLUSIONS

A neighbor network coding scheme for broadcasting in lossy wireless networks is introduced in this chapter. Analytically, network reliability is investigated and reported as exact results, closed-form upper and lower bounds. The framework for analysing network reliability established in this chapter can be extended to investigate network reliability using different coding schemes, for example, each node can choose multiple coding neighbors. Furthermore, it is critical to develop a theoretical proof for the best coding scheme for receiving packets from a specific source node.

1. P. Larsson and N. Johansson, "Multi-user ARQ," in *Proceedings of IEEE Vehicular Technology Conference*, vol. 4, 2006, pp. 2052–2057.

2. F. Kuo, K. Tan, X. Li, J. Zhang and X. Fu, "XOR rescue: exploiting network coding in lossy wireless networks," in *Proceedings of IEEE 6th Annual Communications Society Conference on Sensor, Mesh and Ad Hoc Communications and Networks*, 2009, pp. 1–9.

3. M. Ghaderi, D. Towsley, and J. Kurose, "Network coding performance for reliable multicast," in *Proceedings of IEEE Military Communications Conference*, 2007, pp. 1–7.

4. M. Ghaderi, D. Towsley, and J. Kurose, "Reliability gain of network coding in lossy wireless networks," in *Proceedings of IEEE INFOCOM*, 2008, pp. 2171–2179.

5. M. Nistor, D. Lucani, T. Vinhoza, R. Costa, and J. Barros, "On the delay distribution of random linear network coding," *IEEE Journal on Selected Areas in Communications*, vol. 29, no. 5, pp. 1084–1093, 2011.

6. T. Ho, M. Medard, R. Koetter, D. Karger, M. Effros, S. Jun and B. Leong, "A random linear network coding approach to multicast," *IEEE Transactions on Information Theory*, vol. 52, no. 10, pp. 4413–4460, 2006.

7. G. Mao, and B. Anderson, "Graph theoretic models and tools for the analysis of dynamic wireless multihop networks," in *Proceedings of IEEE Wireless Communications and Networking Conference*, 2009, pp. 1–6.

8 Random Neighbor Network Coding

8.1 INTRODUCTION

A key mechanism in wireless communications is all-to-all broadcast, in which each node has a packet to send to every other node. The packet transmitted from a source node may not be able to reach its destinations in one transmission due to the lossy nature of wireless communications. As a result, multiple transmissions may be required to achieve desired reliability, where reliability is defined as the probability that each node in the network receives or decodes the native packets of every other node.

We present a random neighbor network coding (RNNC) scheme for all-to-all broadcast in lossy wireless networks in this chapter. In the random neighbor network coding scheme, each node chooses at random 1) whether or not to perform coding based on a tuning parameter, and 2) which packets to code on-the-fly based on the packets it has received and decoded. To characterize the reliability of networks using the proposed RNNC scheme, a theoretical analysis is carried out. It is determined which tuning parameter maximizes the reliability of all-to-all broadcast in a network with a given link quality. It is concluded that a node performing coding at every opportunity may not provide optimal results. It is demonstrated that using the proposed RNNC scheme improves network reliability significantly. Furthermore, under $GF(2)$, the reliability performance of RNNC is compared to random linear network coding.

The remainder of the chapter is laid out as follows. The system model is introduced in Section 8.2. Sections 8.3 and 8.4 cover theoretical analysis and optimisation techniques, respectively. The numerical outcomes are then presented in Section 8.5.

8.2 SYSTEM MODEL

The interest model is an all-to-all broadcast in a n node network. It is assumed that time has been set aside. One source node broadcasts a single packet to all other nodes in the network during each time slot, while every other node listens. A successful transmission is not acknowledged because all nodes in the network broadcast in a round-robin fashion. A round, denoted by the letter R, is a set of time slots in which each node broadcasts once. Each node has one packet to broadcast at the start (time 0). Define the packet that a node N_k ($k \in \{1, 2, \ldots, n\}$) has as its native packet at time zero, which is denoted by xk.

Furthermore, it is assumed that a packet sent from N_j will reach N_i in one time slot with a probability of p_{ji}, where $p_{ji} \in (0, 1]$. The p_{ji}s $(i, j \in \{1, 2, \ldots, n\})$ for each pair of nodes are then written as a matrix, which is referred to as the probabilistic connectivity matrix [6]. Because this research focuses on the impact of network coding on reliability, a probabilistic connectivity matrix is assumed to be known.

The packet broadcast from a source node may not be able to reach all destinations in one time slot due to the lossy nature of wireless communications. As a result, retransmission is necessary. A source node typically only re-broadcasts its native packet. A source node can broadcast a coded packet using network coding.

DOI: 10.1201/9781003203803-8

Many existing network coding schemes are fixed network coding schemes, in which network coding is performed on all or a subset of predetermined packets [4,5]. Because lossy channels are common in real-world wireless networks, the packets that a node receives are random, and the predetermined packets may not be available, hence random network coding schemes has recently gained popularity [3].

The RNNC scheme is the network coding scheme used in this chapter. The RNNC scheme's encoding and decoding rules are detailed in the following paragraphs.

ENCODING

Assign $\mathcal{D}_j$ to the collection of native packets that N_j possesses. It contains its own native packets, as well as native packets received directly from other nodes and native packets decoded from coded packets. If $\mathcal{D}_j = \{x_j\}$, then N_j broadcasts x_j. If $\mathcal{D}_j \setminus \{x_j\} \neq \emptyset$, then N_j does not employ network coding and broadcasts the native packet with probability $1 - \omega$; and with probability ω, N_j randomly selects a packet from $\mathcal{D}_j \setminus \{x_j\}$ with equal probabilities and performs bitwise XOR between the selected native packet and its native packet x_j.

The ω parameter was created to address situations where network coding may have a negative impact on reliability. As described in [3], when a node receives a large number of XORed packets, it may be unable to decode them due to a lack of native packets. As a result, even if there are other packets available for coding, ω is used to allow a node to choose to broadcast its native packet with a certain probability. If $\mathcal{D}_2 = \{x_1, x_2, x_3\}$, for example, N_2 broadcasts x_2, $x_1 \oplus x_2$ and $x_2 \oplus x_3$ with probabilities of $1 - \omega$, $\omega/2$, and $\omega/2$, respectively. In addition, in Section 8.4, the optimal value of ω that maximizes reliability will be investigated.

DECODING

The successful decoding in a coded packet requires only the successful receipt or decoding of one of the 2 native packets that make up the coded packet. x_z, for example, can be decoded from packets $x_z \oplus x_k$ and x_k by performing $(x_z \oplus x_k) \oplus x_k$, where $z, k \in \{1, 2..., n\}$ and $y \neq k$. Please note that x_k can be either directly received or decoded from another encoded packet. Used to decode other coded packets after x_z and x_k has been decoded. Until XORed packets can be decoded, the decoding process continues.

Each node utilizes a buffer to store packets, including native packets, packets received from other nodes, and decoded packets. Packets that are duplicated are dropped. Additionally, if the buffer already contains all native packets that comprise a coded packet, the coded packet is dropped. This way, the demand for the buffer's size can be minimized.

8.3 THEORETICAL ANALYSIS

A source node broadcasts a packet during each time slot. This packet may contain different information depending on the packets stored in the source node, according to the encoding rules. Additionally, the packet may be received with varying probabilities by different destination nodes. As a result, tracking the packets that each node receives and stores following each transmission is challenging.

Let $\Upsilon_k(t)$ represent the packets that N_k has at the end of the time slot t. Consequently, denote the packets stored at all the nodes in the network at the end of time slot t as $\boldsymbol{v}(t) = [\boldsymbol{v_1}(t); \boldsymbol{v_2}(t); \ldots; \boldsymbol{v_n}(t)]$. Let random processes $\mathcal{Z}_k(t)$ and $\mathcal{Z}(t)$ represent packets of an individual node N_k stored in its buffer at the end of time slot t, and the packets that are individually stored at every node in the network at the end of time slot t respectively. The

Random Neighbor Network Coding

packets received and stored in the previous 0 to $t-1$ time slots are included in the packets stored at a node in time slot t. With the definition of $\mathcal{Z}_k(t)$, we have:

$$\Pr\Big(\mathcal{Z}_k(t+1) = \boldsymbol{v_k}(t+1) \mid \mathcal{Z}_k(0) = \boldsymbol{v_k}(0), \mathcal{Z}_k(1) = \boldsymbol{v_k}(1), \ldots, \mathcal{Z}_k(t) = \boldsymbol{v_k}(t)\Big)$$
$$= \Pr\Big(\mathcal{Z}_k(t+1) = \boldsymbol{v_k}(t+1) \mid \mathcal{Z}_k(t) = \boldsymbol{v_k}(t)\Big). \tag{8.1}$$

Similarly, the random process that governs the received packets of every node in the network in the time slot $t+1$ depends only on the received packets of every node in the time slot t, but not on those before time slot t. There is:

$$\Pr\Big(\mathcal{Z}(t+1) = \boldsymbol{\Upsilon}(t+1) \mid \mathcal{Z}(1) = \boldsymbol{\Upsilon}(1), \mathcal{Z}(2) = \boldsymbol{\Upsilon}(2), \ldots, \mathcal{Z}(t) = \boldsymbol{\Upsilon}(t)\Big)$$
$$= \Pr\Big(\mathcal{Z}(t+1) = \boldsymbol{\Upsilon}(t+1) \mid \mathcal{Z}(t) = \boldsymbol{\Upsilon}(t)\Big). \tag{8.2}$$

It is clear that these processes are memoryless. Further, the probability $\Pr(\mathcal{Z}(t+1) = \boldsymbol{\Upsilon}(t+1))$ can be obtained from the probabilities $\Pr(\mathcal{Z}_k(t+1) = \boldsymbol{v_k}(t+1))$ for every $k \in \{1, 2, .., n\}$. More specifically, $\Pr(\mathcal{Z}(t+1))$ is the product of every $\Pr(\mathcal{Z}_k(t+1))$ for $k \in \{1, 2, .., n\}$. By examining the change in packets at each individual node in the network after each node broadcasts, a Markov chain that governs the random process $\mathcal{Z}$ can be established. The Markov chain can be used to investigate network reliability.

8.3.1 STATES

The ath state of a network is represented by the matrix $\boldsymbol{\Upsilon}_a$, where the state refers to the status of stored packets at all nodes and $a \in \{1, 2, \ldots, L\}$, where L is the total number of states. Let the packets that node $\boldsymbol{N_k}$ has when the network is in state $\boldsymbol{\Upsilon}_a$ be denoted by $\boldsymbol{v_{ka}}$ in the kth row of $\boldsymbol{\Upsilon}_a$.

In the network, there are two types of packets. The native packet of each source node falls into the first category, while the XORed packet of a pair of native packets falls into the second. A unique index is assigned to each packet. The native packet $\boldsymbol{x_k}$ is assigned index k, while an XORed packet, say $\boldsymbol{x_\gamma} \oplus \boldsymbol{x_k}$ (it is assumed $1 \leq \gamma < k \leq n$ without sacrificing generality), is assigned index:

$$\mu_{\gamma,k} \triangleq n + [(n-1) + (n-\gamma+1)](\gamma-1)/2 + (k-\gamma)$$
$$= n\gamma - \gamma^2/2 - \gamma/2 + k. \tag{8.3}$$

Then, the total number of distinct packets is:

$$n + \binom{n}{2} = \frac{(n^2+n)}{2}, \tag{8.4}$$

where n represents the total number of distinct native packets and $\binom{n}{2}$ represents the total number of distinct XORed packets As a result, $\boldsymbol{v_{ka}}$ can be represented as a row vector made up of $(n^2+n)/2$ elements, each of which represents a packet. Specifically, the first n elements of $\boldsymbol{v_{ka}}$ represent native packets of n nodes, while the subsequent elements represent

XORed packets. When node N_k has possession of a packet, the corresponding element in v_{ka} is set to one; otherwise, it is set to zero. If node N_1 has packets x_1 and $x_2 \oplus x_3$ in a network with three nodes, then $v_{1a} = [1, 0, 0, 0, 0, 1]$.

Absorbing state

There is an absorbing state that represents the event when all nodes in the network have successfully received or decoded each other's native packets. When this happens, the status of encoded packets becomes irrelevant. The state Υ_L, in which the first n elements in each row are one and the rest are zero, represents this state.

States reduction

Two methods for reducing the number of states are introduced to reduce the complexity of the analysis. To begin, the number of states can be reduced by considering the decoding process, i.e., states containing XORed packets whose native packets have already been received/decoded are merged.

Second, invalid states can be discarded to further reduce the number of states. The invalid states are those that cannot be entered at any time. When neither N_γ nor N_k has both x_γ and x_k, a third node N_θ with $\theta \in \{1, 2, \ldots, n\} \setminus \{\gamma, k\}$ cannot have encoded packet $x_\gamma \oplus x_k$. As a result, the associated state is null.

Remark 1. *The total number of states for a network with n nodes can be calculated before reduction by:*

$$2^{(\frac{n^2+n}{2}-1) \times n} = 2^{\frac{n^3+n^2-2n}{2}} \tag{8.5}$$

Because each v_{ka} (for $k \in \{1, 2, ldots, n\}$) has $n^2/2 + n/2 - 1$ bits that can take the value of 0 or 1. The minus one in the expression denotes that one bit is set to a specific value. This is because node N_k always has x_k and the kth bit of v_{ka} is always one. The number of states could be reduced significantly as a result of the state reduction.

8.3.2 TRANSITION MATRICES

The packet broadcast from a source node may be received by destination nodes with varying probabilities at the end of each time slot, resulting in an update of the status of packets stored in some nodes. A transition of states in the Markov chain, which was used to describe the packets stored in each node of the network, can reflect this.

Fig. 8.1 shows some states of a network's Markov chain when $n = 3$ to demonstrate this. In Fig. 8.1, Υ_1 represents the initial state, Υ_L represents the absorbing state, and $\zeta_j(\Upsilon_A|\Upsilon_B)$ represents the transition probability of the network changing from state Υ_B to state Υ_A when N_j broadcasts.

When each node broadcasts, the transition matrices are constructed separately in this section. When N_j broadcasts, denote $\mathcal{M}_j$ as the transition matrix. It's a $L \times L$ matrix, which is defined as follows:

$$\mathcal{M}_j = \begin{bmatrix} \zeta_j(\Upsilon_1|\Upsilon_1) & \cdots & \zeta_j(\Upsilon_L|\Upsilon_1) \\ \vdots & \ddots & \vdots \\ \zeta_j(\Upsilon_1|\Upsilon_L) & \cdots & \zeta_j(\Upsilon_L|\Upsilon_L) \end{bmatrix} \tag{8.6}$$

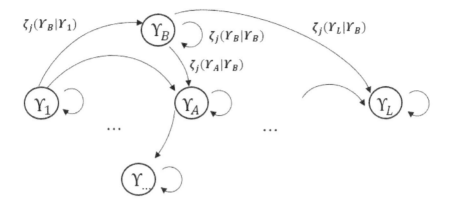

FIGURE 8.1: A state transition diagram for a wireless network with three nodes.

where the entry in row **B** and column **A**, denoted by $\zeta_j(\Upsilon_A|\Upsilon_B)$, represents the probability that the network state changes from Υ_B in one time slot to Υ_A in the next time slot. The following content shows how to calculate an entry $\zeta_j(\Upsilon_A|\Upsilon_B)$.

The first step is to look at the packets that were stored in the transmitting node N_j during these time slots, which are v_{jA} and v_{jB}. If $v_{jA} \neq v_{jB}$, then the transition probability $\zeta_j(\Upsilon_A|\Upsilon_B) = 0$. This is due to the fact that the transmitting node N_j only broadcasts packets that it already has and will not cause a variation of its own packets.

After that, we consider the case where $v_{jA} = v_{jB}$. Keep in mind that N_j's native packets are $\mathcal{D}_j$. The cardinality of $\mathcal{D}_j$ is denoted by m_j. The source node has m_j different packets for transmission, including $m_j - 1$ XORed packets and one native packet. π_{jh} is the event in which N_j broadcasts an XOR coded packet $x_j \oplus x_h$, and π_{jj} is the event in which N_j broadcasts its native packet x_j, where $h \in \mathcal{D}_j \setminus \{j\}$. Based on the rules for RNNC mentioned in Section 8.2, we have

$$\Pr(\pi_{jh}) = \frac{\omega}{m_j - 1},$$
$$\Pr(\pi_{jj}) = 1 - \omega. \qquad (8.7)$$

Moreover, denote $\Pr(\Upsilon_A|\Upsilon_B, \pi_{jj})$ and $\Pr(\Upsilon_A|\Upsilon_B, \pi_{jh})$ as the conditional probabilities that the network state changes from Υ_B to Υ_A if the source node broadcasts packets x_j and $x_j \oplus x_h$, respectively. Then based on the total probability theory [7], we have,

$$\zeta_j(\Upsilon_A|\Upsilon_B) = \Pr(\Upsilon_A|\Upsilon_B, \pi_{jj}) \Pr(\pi_{jj}) + \sum_{h \in \mathcal{D}_j \setminus \{j\}} \Pr(\Upsilon_A|\Upsilon_B, \pi_{jh}) \Pr(\pi_{jh})$$
$$= \Pr(\Upsilon_A|\Upsilon_B, \pi_{jj})(1 - \omega) + \sum_{h \in \mathcal{D}_j \setminus \{j\}} \Pr(\Upsilon_A|\Upsilon_B, \pi_{jh}) \frac{\omega}{m_j - 1}.$$
$$(8.8)$$

The transition of state from Υ_B to Υ_A, contains n independent events which the status of the packets in the buffer of node k changes from Υ_{kB} to v_{kA} for all $k \in \{1, 2 \ldots, n\}$.

Denoted by $P_{ji}^{h}(v_{iA}|v_{iB})$ the probability that the status of the packets in the buffer of node N_i changes from v_{iB} to v_{iA} for all $i \in \{1,2\ldots,n\}$ when node N_j transmits $x_j \oplus x_h$. Then, denote by $P_{ji}^{j}(v_{iA}|v_{iB})$ the probability that the status of packets in node N_i change from v_{iB} to v_{iA} for all $i \in \{1,2\ldots,n\}$ when node N_j transmits x_j.

It is obvious that the conditional probability $\Pr(\Upsilon_A|\Upsilon_B, \pi_{jh})$ is the product of n probabilities $P_{ji}^{h}(v_{iA}|v_{iB})$. Similarly, the conditional probability $\Pr(\Upsilon_A|\Upsilon_B, \pi_{jj})$ can be calculated by the product of n probabilities $P_{ji}^{j}(v_{iA}|v_{iB})$ for all $i \in \{1,2,\ldots,n\}$. There are:

$$\Pr(\Upsilon_A|\Upsilon_B, \pi_{jj}) = \prod_{i \in \{1,2,\ldots,n\}} P_{ji}^{j}(v_{iA}|v_{iB}),$$

$$\Pr(\Upsilon_A|\Upsilon_B, \pi_{jh}) = \prod_{i \in \{1,2,\ldots,n\}} P_{ji}^{h}(v_{iA}|v_{iB}). \quad (8.9)$$

$P_{ji}^{j}(v_{iA}|v_{iB})$ and $P_{ji}^{h}(v_{iA}|v_{iB})$ are calculated by using Algorithms 8.1 and 8.2, respectively.

In algorithms 8.1 and 8.2, we let $v_{iA}\{\ell\}$ (resp. $v_{iB}\{\ell\}$) represent the ℓth element of v_{iA} (resp. v_{iB}) and $v_{iA} = v_{iB}$ if $v_{iA}\{\ell\} = v_{iB}\{\ell\}$ for all $\ell \in \{1,2,\ldots,(n^2+n)/2\}$.

Algorithm 8.1 when N_j transmits x_j to N_i
---

if $v_{iB}\{j\} = 1$ then
    if $v_{iB} = v_{iA}$ then $P_{ji}^{j}(v_{iA}|v_{iB}) = 1$;
    else $P_{ji}^{j}(v_{iA}|v_{iB}) = 0$;
    end if
else
    if $v_{iB}\{\mu_{j,\lambda}\} = 1$ for any $\lambda \in \mathcal{N} \setminus \{i,j\}$ then
        if $v_{iA}\{x\} = 1$ and $v_{iA}\{y\} = 0$ for all $x \in \mathcal{H}_{ji}$, $y \in \mathcal{G}_{ji}$; and $v_{iB}\{\lambda\} = v_{iA}\{\lambda\}$ for all $\lambda \in \{1,2,\ldots(n^2+n)/2\} \setminus \{\mathcal{H}_{ji}, \mathcal{G}_{ji}\}$ then $P_{ji}^{j}(v_{iA}|v_{iB}) = p_{ji}$;
        else if $v_{iB} = v_{iA}$ then $P_{ji}^{j}(v_{iA}|v_{iB}) = 1 - p_{ji}$;
        else $P_{ji}^{j}(v_{iA}|v_{iB}) = 0$;
        end if
    else
        if $v_{iA}\{j\} = 1$ and $v_{iB}\{\lambda\} = v_{iA}\{\lambda\}$ for all $\lambda \in \{1,2,\ldots,(n^2+n)/2\} \setminus \{i\}$ then $P_{ji}^{j}(v_{iA}|v_{iB}) = p_{ji}$;
        else if $v_{iB} = v_{iA}$ then $P_{ji}^{j}(v_{iA}|v_{iB}) = 1 - p_{ji}$;
        else $P_{ji}^{j}(v_{iA}|v_{iB}) = 0$.
        end if
    end if
end if

In Algorithm 8.1, $\mathcal{H}_{ji}$ is the set of indices of the native packets that upon receiving packet x_j node N_i is able to decode. For example, if N_i has the XORed packet $(x_\gamma \oplus x_k)$ and $\mathcal{H}_{ji}$ has the index of one of the native packets (either γ or k), then the indices k or γ will be added in $\mathcal{H}_{ji}$. $\mathcal{G}_{ji}$ is the set of the indices of the XORed packets each of which can be decoded by using Eq. (8.3) at node N_i when the packet x_j is received.

Algorithm 8.2 when N_j transmit $x_j \oplus x_h$ to N_i for all $h \in \mathcal{D}_i$

if $v_{iB}\{\mu_{j,h}\} = 1$ or ($v_{iB}\{j\} = 1$ and $v_{iB}\{h\} = 1$) then
  if $v_{iB} = v_{iA}$ then $P_{ji}^h(v_{iA}|v_{iB}) = 1$;
  else $P_{ji}^h(v_{iA}|v_{iB}) = 0$;
  end if
else if $v_{iB}\{j\} = 0$ and $v_{iB}\{h\} = 0$ then
  if $v_{iA}\{\mu_{j,h}\} = 1$ and $v_{iB}\{\lambda\} = v_{iA}\{\lambda\}$, for all $\lambda \in \{1, 2, \ldots, (n^2 + n)/2\} \setminus \{\mu_{j,\lambda}\}$ then $P_{ji}^h(v_{iA}|v_{iB}) = p_{ji}$;
  else if $v_{iB} = v_{iA}$ then $P_{ji}^h(v_{iA}|v_{iB}) = 1 - p_{ji}$;
  else $P_{ji}^h(v_{iA}|v_{iB}) = 0$;
  end if
else
  if $v_{iA}\{x\} = 1$ and $v_{iA}\{y\} = 0$ for all $x \in \mathcal{H}_{jhi}, y \in \mathcal{G}_{jhi}$; and $v_{iB}\{\lambda\} = v_{iA}\{\lambda\}$ for all $\lambda \in \{1, 2, \ldots, (n^2 + n)/2\} \setminus \{\mathcal{H}_{jhi}, \mathcal{G}_{jhi}\}$ then $P_{ji}^h(v_{iA}|v_{iB}) = p_{ji}$;
  else if $v_{iB} = v_{iA}$ then $P_{ji}^h(v_{iA}|v_{iB}) = 1 - p_{ji}$;
  else $P_{ji}^h(v_{iA}|v_{iB}) = 0$;
  end if
end if

In Algorithm 8.2, $\mathcal{H}_{jhi}$ is the set of indices of the native packets that upon receiving packet $x_j \oplus x_h$, node N_i is able to decode. $\mathcal{G}_{jhi}$ the set of indices of all XORed packets that can be decoded at node N_i upon receiving packet $x_j \oplus x_h$.

After obtaining every $P_{ji}^j(v_{iA}|v_{iB})$ and $P_{ji}^h(v_{iA}|v_{iB})$, substitute equation (8.9) to equation (8.8). The entry $\zeta_j(\Upsilon_A|\Upsilon_B)$ is readily obtained, which is:

$$\zeta_j(\Upsilon_A|\Upsilon_B) = \Pr(\Upsilon_A|\Upsilon_B, \pi_{jj})(1-\omega) + \sum_{h \in \mathcal{D}_j \setminus \{j\}} \Pr(\Upsilon_A|\Upsilon_B, \pi_{jh}) \frac{\omega}{m_j - 1}. \quad (8.10)$$

The first and the second term of right hand side of Eq. 8.10 are given by

$$\prod_{i \in \{1,2,\ldots,n\}} P_{ji}^j(v_{iA}|v_{iB})(1-\omega) \quad (8.11)$$

and
$$\sum_{h \in \mathcal{D}_j \setminus \{j\}} \prod_{i \in \{1,2,\ldots,n\}} P_{ji}^h \left(v_{iA} | v_{iB} \right) \frac{\omega}{m_j - 1}, \tag{8.12}$$

respectively.

The first row of the transition matrix $\mathcal{M}_j$ is obtained by setting the state Υ_B to Υ_1 and varying the state Υ_A from Υ_1 to Υ_L. Then repeat the procedure for each Υ_B, where $B \in \{2, 3, \ldots, L\}$, to obtain the transition matrix $\mathcal{M}_j$, which governs the network's state transition when N_j transmits. Similarly, for each $j \in \{1, 2, \ldots, n\}$, the transition matrices $\mathcal{M}_j$ can be calculated. As a result, define the transition matrix for a round in which each node broadcasts once as $\mathcal{M} \triangleq \prod_{j=1}^n \mathcal{M}_j$.

It's worth noting that the transition matrices' entries are found by taking the product of the tuning parameter and a series of p_{ji} and $1 - p_{ji}$, where $i, j \in \{1, 2, \ldots, n\}$. p_{ji} is the corresponding entry of a probabilistic connectivity matrix. As a result, if the probabilistic connectivity matrix is the same during the time slots when N_j transmits, the $\mathcal{M}_j$ during these time slots is the same.

8.3.3 PROBABILITY VECTOR AND THE RELIABILITY

The probability vector represents the network's probabilities in each possible state. It's a $1 \times L$ row vector, with the lth element representing the probability that the network is in state Υ_l. The probability vector at the end of each round, R, is denoted by the symbol $S(R)$. The element of the initial state is of probability one, and all other elements are zero, in the initial probability vector, denoted by $S(0)$, where the initial state is that every node only has its native packet, denoted by Υ_1. The probability vector at the end of round R, according to Markov theory, is equal to:

$$S(R) = S(R-1) \prod_{j=1}^n \mathcal{M}_j$$
$$= S(0)(\prod_{j=1}^n \mathcal{M}_j)^R$$
$$= S(0)\mathcal{M}^R. \tag{8.13}$$

Finally, at the end of round R, we calculate the reliability, which is denoted by $\psi(R)$. It's the probability that by the end of round R, the network will be in an absorbing state.

$$\psi(R) = S(R)\{L\}, \tag{8.14}$$

where the absorbing state is indicated by the Lth bit in the probability vector.

8.4 OPTIMISATIONS

The encoding at a source node N_j in a network using the random neighbor network coding scheme is done on a randomly selected packet from the set $\mathcal{D}_j \setminus \{x_j\}$, where $\mathcal{D}_j$ is defined in Section 8.2. Furthermore, the received packets at a node are influenced by the probabilistic connectivity matrix, which is a network property that cannot be changed. The tuning parameter ω, on the other hand, which is the probability of a source node performing network coding, can be controlled.

Random Neighbor Network Coding

The ω value range is $[0, 1]$, where 0 indicates that a node does not perform coding and 1 indicates that a node always performs coding. As a result, the impact of coding on reliability is determined by this parameter. We can tune the reliability of a network ranging from non-coding to absolutely coding by adjusting ω. When a network's probabilistic connectivity matrix is given, it's critical to find the optimal ω that maximizes network reliability.

8.4.1 OPTIMIZE THE RELIABILITY AT AN INDIVIDUAL ROUND

The network reliability $\psi(R)$ at round R, given in equation (8.14), is a function of variables ω, p_{ji} and $1 - p_{ji}$ where $i, j \in \{1, 2, \ldots, n\}$. In the case that the probabilistic connectivity matrix, denoted by Q, is given, the expression $\psi(R)$ can be reduced to a single-variable polynomial by substituting the values of the given Q. There is:

$$\psi(R) = f(\omega), \tag{8.15}$$

where $\omega \in [0, 1]$. It is a constrained nonlinear optimisation problem. Detailed methods will be given in Section 8.5.

8.4.2 OPTIMIZE THE EXPECTED ROUND TO ABSORB

Remember that the Markov chain has only one absorbing state. Rearrange the transition matrix in the following order to achieve the canonical form:

$$\mathcal{M} = \begin{pmatrix} W & Y \\ 0 & 1 \end{pmatrix}, \tag{8.16}$$

where W denotes transitions between all transient states and Y denotes transitions from transient to absorbing states. The fundamental matrix, denoted by N, is as follows:

$$N = (I - W)^{-1}, \tag{8.17}$$

where the $(A, B)^{th}$ entry denotes the number of rounds required to reach a transient state B from a transient state A. Finally, the expected number of rounds to reach the absorbing state, in which each node receives or decodes all other nodes' native packets, can be calculated by:

$$E = Nc, \tag{8.18}$$

where c is a one-dimensional column vector with a size of $L \times 1$. As a result, the entry $E\{1\} \triangleq E_{exp}$, which is a function of ω, p_{ji} and $1 - p_{ji}$, is the expected number of rounds to reach absorbing state Υ_L (assume it is the last state) from initial state Υ_1 (assume it is the first state). E_{exp} becomes a single variable polynomial of ω when the values of a given probabilistic connectivity matrix are substituted. Finally, a constrained nonlinear optimisation can easily find the minimum E_{exp} and the corresponding ω.

8.5 NUMERICAL RESULTS

This section provides numerical evaluations of the network reliability analytical results obtained in this chapter. There are also examples of procedures for optimising the random neighbor network coding scheme. The proposed scheme's reliability gain is then tested in networks with various parameters.

Remember that arbitrary values in the probabilistic connectivity matrices indicating the end-to-end connection probabilities of pairs of nodes are allowed in the analysis. Their entries are generated at random in this section by uniformly selecting from $(0, 1]$ for numerical evaluations.

8.5.1 VALIDATION OF THE THEORETICAL ANALYSIS

Random neighbor network coding is used on networks with three or four nodes. Fig. 8.2 shows the analytical results of reliability after each round of transmissions, which were obtained using the methods described in Section 8.3. For $n = 3$ and $n = 4$, the probabilistic connectivity matrices used in Fig.8.2 are:

$$Q_3 = \begin{bmatrix} 1 & 0.2 & 0.3 \\ 0.4 & 1 & 0.5 \\ 0.6 & 0.7 & 1 \end{bmatrix}; Q_4 = \begin{bmatrix} 1 & 0.2 & 0.3 & 0.6 \\ 0.4 & 1 & 0.5 & 0.3 \\ 0.6 & 0.7 & 1 & 0.2 \\ 0.3 & 0.4 & 0.5 & 1 \end{bmatrix}, \qquad (8.19)$$

and ω is set to 0.6 (note that the optimal value of ω is determined in the following subsection). The simulation results are also plotted for comparison under the same network configurations. The Monte Carlo simulation results are averaged values from multiple 10^5 runs. The theoretical results closely match the simulation results, as shown in Fig. 8.2, validating the theoretical analysis.

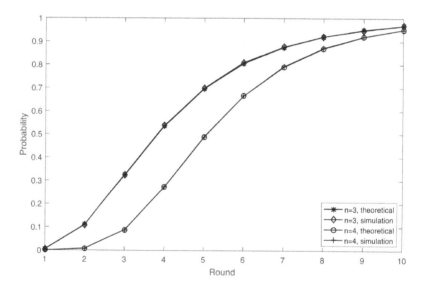

FIGURE 8.2: Theoretical and simulation results comparison for the network reliability applying RNNC when n = 3 and 4. The probabilistic connectivity matrices are given by Eq. (8.19).

8.5.2 OPTIMAL SELECTION OF THE TUNING PARAMETER

Two optimizations are performed in the following: 1) maximize reliability at a specific round, or 2) reduce the expected number of rounds to reach the absorbing state.

Consider a network with three nodes and the following probabilistic connectivity matrix:

$$Q'_3 = \begin{bmatrix} 1 & 0.2 & 0.3 \\ 0.4 & 1 & 0.5 \\ 0.6 & 0.7 & 1 \end{bmatrix}. \quad (8.20)$$

The network's reliability at round $R = 4$ can be calculated using the $\psi(4)$ Eq. (8.14). The reliability expression is then simplified into a polynomial of a single variable ω (rounded to four decimal places) by substituting the corresponding entries of the probabilistic connectivity matrix into $\psi(4)$:

$$\psi(4) = -0.0003\omega^6 + 0.0083\omega^5 - 0.0468\omega^4 \\ + 0.1362\omega^3 - 0.4154\omega^2 + 0.1285\omega + 0.5432, \quad (8.21)$$

where $\omega \in [0, 1]$.

Taking differentiation:

$$\frac{d(\psi(4))}{d\omega} = -0.0018\omega^5 + 0.0415\omega^4 - 0.1872\omega^3 \\ + 0.4086\omega^2 - 0.8308\omega + 0.1285. \quad (8.22)$$

If $\frac{d(\psi(4))}{d\omega} = 0$, then there exists a solution for ω, subject to the constraint that $\omega \in [0, 1]$, which also gives a maximum value for $\psi(4)$, where $\psi(4) = 0.5537$ and $\omega = 0.8325$. The reliability at the end of round $R = 4$ is plotted against ω in Fig. 8.3. The optimal ω value that maximizes network reliability is highlighted.

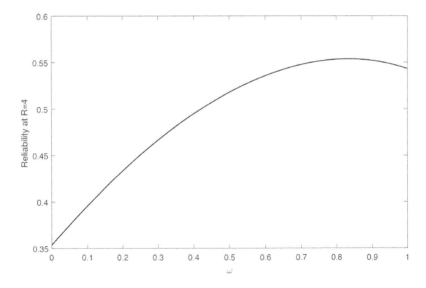

FIGURE 8.3: The reliability for a three node network at the 4th round when ω varies form zero to one, the probabilistic connectivity matrix is given by eq. (8.20).

TABLE 8.1
Optimal ω values for a three node network at different round with the probabilistic connectivity matrix given by Eq. 8.20.

Round	3	4	5	6	7	8
optimal ω	0.8476	0.8325	0.8310	0.8349	0.8408	0.8472

The network reliability at the end of round $R = 6$ is shown in Fig. 8.4, which is calculated using the same method. The curve shows that $\psi(6)$ and $\psi(4)$ have a similar tendency. Furthermore, $\psi(6)$ has a maximum value of 0.8243, which occurs when $\omega = 0.8349$. Table 8.1 shows the optimized tuning parameters that maximize network reliability from $R = 3$ to 8.

Furthermore, both the minimum values for $\psi(4)$ and $\psi(6)$ occur when $\omega = 0$, as shown in Fig. 8.3 and Fig. 8.4. Because $\omega = 0$ corresponds to all nodes transmitting only their native packets, it follows that when $\omega > 0$, networks using the proposed coding scheme always outperform non-coded networks with the same setting. From a theoretical standpoint, this conclusion is simple to prove.

Then, using the methods described in Section 8.4.2, optimisation is carried out with the goal of reducing the expected number of rounds required to reach the absorbing state. Similarly, E_{exp} is converted to a single variable polynomial of ω after substituting values from Q. Then, by solving a constrained nonlinear optimisation problem, the minimum value of E_{exp} and the

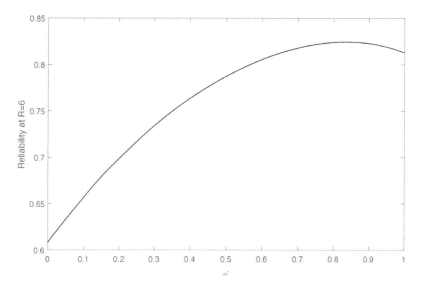

FIGURE 8.4: The reliability for a three node network at the 6th round when ω varies form zero to one, the probabilistic connectivity matrix is given by eq. (8.20).

Random Neighbor Network Coding

corresponding ω can be found, which is similar to the previous optimisation problem, so the details are omitted here. When $\omega = 0.8460$, the minimum expected number of rounds to reach the absorbing state is 4.7147.

As shown in Figs. 8.3 and 8.4, the optimal tuning parameter ω that maximizes reliability at round $R = 4$ and $R = 6$ is 0.8325 and 0.8349, respectively. Furthermore, 0.8460 is the optimal ω for reducing the expected number of rounds to the absorbing state. The optimal ω for the expected number of rounds to reach the absorbing state may not be optimal for the reliability at each individual round, as can be seen. These optimal ω values, on the other hand, are close to each other, as shown in Table 8.1. Round $R = 3$ to $R = 8$, the expected value for the optimal ω is 0.8390, with a variance of 0.0164. As a result, the network designer can select the best ω value for a particular round without jeopardising the reliability of subsequent rounds.

8.5.3 EXAMINATION ON THE RELIABILITY GAIN

The proposed scheme is applied to networks with various configurations in order to examine the gain in reliability that random neighbor network coding brings. The proposed scheme is shown in Fig. 8.5 for networks with three and four nodes. The tuning parameters are set to the optimal values for the fourth round, which are 0.833 and 0.965 for $n = 3$ and 4, respectively. For $n = 3$ and $n = 4$, the probabilistic matrices are as follows:

$$Q_3'' = \begin{bmatrix} 1 & 0.2 & 0.3 \\ 0.4 & 1 & 0.5 \\ 0.6 & 0.7 & 1 \end{bmatrix} ; Q_4'' = \begin{bmatrix} 1 & 0.1 & 0.5 & 0.4 \\ 0.6 & 1 & 0.2 & 0.6 \\ 0.7 & 0.3 & 1 & 0.1 \\ 0.1 & 0.3 & 0.2 & 1 \end{bmatrix}. \quad (8.23)$$

For comparison, we also plotted the reliabilities for non-coded networks.

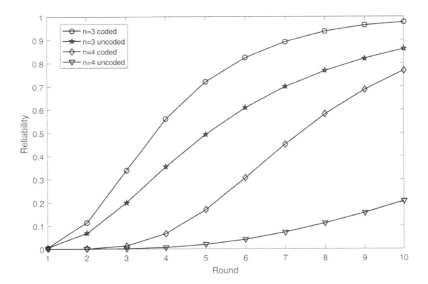

FIGURE 8.5: The reliability comparison for networks with three and four nodes. The connectivity matrices are given in eq. (8.23).

The reliability of the networks using the proposed random neighbor network coding scheme outperforms the corresponding non-coded networks, as shown in Fig. 8.5. Furthermore, in some cases, the increase in reliability is significant. As shown in Fig. 8.5, the reliability of a coded network with four nodes at round $R = 10$ is 0.7685, while for the corresponding non-coded network it is 0.2064.

8.5.4 COMPARISON WITH THE RANDOM LINEAR NETWORK CODING SCHEME

Because it is one of the most popular network coding schemes, the random linear network coding scheme [8] was chosen as the benchmark. The reliability performance of a network using a random linear network coding scheme is compared to that of the proposed schemes in this section. The finite field is set to $GF(2)$ to ensure a fair comparison. Then, the RNNC becomes the XOR coding on randomly selected incoming information flow, so that its computational complexity and bandwidth consumption of a overhead packet is similar to the proposed neighbor coding schemes.

Within one round of simulations, a source node performs random linear network coding on all received packets from other nodes, where each received packet is a linear combination of some native packets. The randomly generated coefficients for each received packet cannot all equal zero at the same time in one encoding procedure, where the zero coefficient for a received packet indicates that this received packet will not be included in the encoding of the transmitted packet. As a result, instead of transmitting nothing, each node must broadcast a packet.

In Fig. 8.6, the reliability of a four-node network applying the proposed two schemes, random linear network coding, and no coding are plotted. The probabilistic connectivity

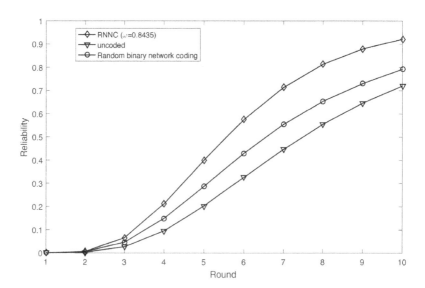

FIGURE 8.6: Performance comparison for different network coding schemes. The probabilistic connectivity matrix is given by Eq. (8.24).

matrix of the network is:

$$Q_4''' = \begin{bmatrix} 1 & 0.3 & 0.5 & 0.4 \\ 0.6 & 1 & 0.2 & 0.6 \\ 0.7 & 0.3 & 1 & 0.4 \\ 0.5 & 0.3 & 0.2 & 1 \end{bmatrix}. \quad (8.24)$$

Furthermore, the tuning parameter in the random neighbor network coding scheme is set to the optimal value that maximizes network reliability at round $R = 4$, where $\omega = 0.8435$.

It is shown that in the optimal scenarios, the proposed schemes outperform the random linear network coding scheme where finite filed for the coefficients is $GF(2)$. It is concluded that the proposed scheme can be optimized and better fit in networks with given probabilistic connectivity matrices, while the random linear network coding scheme under $GF(2)$ can only bring moderate amount of gain in reliability.

8.6 SUMMARY

This chapter describes a random neighbor network coding scheme that allows a source node to perform on-the-fly network coding based on the packets it receives. The reliability of networks using the investigated coding scheme is examined, and the reliability improvement is demonstrated. The optimal tuning parameter that maximizes reliability at a given round or reduces the expected number of rounds to reach the absorbing state has also been found. Finally, in $GF(2)$, the proposed scheme is compared to non-coding and random linear network coding. It is concluded that a node performing coding at every opportunity may not provide optimal results.

1. S. Katti, H. Rahul, H. Wenjun, D. Katabi, M. Medard, and J. Crowcroft, "Xors in the air: Practical wireless network coding," *IEEE/ACM Transactions on Networking*, vol. 16, no. 3, pp. 497–510, 2008.

2. N. Dong, T. Tuan, N. Thinh, and B. Bose, "Wireless broadcast using network coding," *IEEE Transactions on Vehicular Technology*, vol. 58, no. 2, pp. 914–925, 2009.

3. W. Zhe and M. Hassan, "Blind xor: Low-overhead loss recovery for vehicular safety communications," *IEEE Transactions on Vehicular Technology*, vol. 61, no. 1, pp. 35–45, 2012.

4. M. Ghaderi, D. Towsley, and J. Kurose, "Reliability gain of network coding in lossy wireless networks," in *Proceedings of IEEE INFOCOM*, 2008, pp. 2171–2179.

5. M. Nistor, D. E. Lucani, T. T. V. Vinhoza, R. A. Costa, and J. Barros, "On the delay distribution of random linear network coding," *IEEE Journal on Selected Areas in Communications*, vol. 29, no. 5, pp. 1084–1093, 2011.

6. S. Dasgupta and G. Mao, "On the quality of wireless network connectivity," in *Proceedings of IEEE Globecom*, pp. 518–523, 2012.

7. S. W. J., *Probability, Markov chains, queues, and simulation : the mathematical basis of performance modeling.*   Princeton, N.J. : Princeton University Press, 2009.

8. T. Ho, M. Medard, R. Koetter, D. R. Karger, M. Effros, S. Jun, and B. Leong, "A random linear network coding approach to multicast," *Information Theory, IEEE Transactions on*, vol. 52, no. 10, pp. 4413–4430, 2006.

Index

99% bandwidth 61

Access point, 11, 113
Adaptive modulation and coding, 45
All-to-all, 10, 113, 127
Amplify-and-forward, 3, 15, 103
ARQ 9, 11, 13, 49, 50
Average latency, 10
AWGN channel, 51, 59, 60, 67, 69

Bandwidth and power efficiency, 6
Bandwidth efficiency, 1, 6, 7, 8, 103
Bandwidth-efficient, 7
Butterfly, 7

Channel coding, 1, 47, 62
Channel corruption, 28
Channel state information, 33, 106
Check node, 91, 106
Circular, 9
Closed-form, 10, 16, 114, 119, 124
Coded modulation, 15, 47, 48, 51, 53
Codeword error rate, 15, 67, 82
Coding-aware routing, 8
Coding-oblivious routing, 8
Combination networks, 8
Complex baseband, 51, 52
Compute-and-forward, 2, 71, 82, 87, 100
Conflict-free disjoint subnetworks, 25
Continuous Phase Frequency Shift Keying, 51
Crossover probability, 35, 36, 76
Cumulative Distribution Function, 29
Cutoff rate, 40

Decentralized, 6, 14
Decode-and-forward, 3, 15, 103
Deep fading, 40, 110
Deinterleaver, 57
Distributed coding, 47
Distributed network-channel codes, 47
Dither vector, 70, 86

Edge-wise, 9
End-to-end, 8, 135
Energy efficiency, 8, 9, 12

Error-free multicast, 6
Estimate-and-forward, 3, 15, 103
EXIT chart, 58, 60, 61, 62

Fairness requirements, 12
Feedback, 4, 10, 12, 103
Forward Error Correction, 13, 113
Free distance, 36, 38

Gaussian random process, 51, 52
Generator matrices, 26, 37, 38, 39, 44
Generator polynomial, 37, 49, 59
Graph, 4, 9, 90

Hypercube, 75, 77, 92, 97
Hypergraph, 26, 28

Incremental redundancy, 9, 49
Interference-aware routing, 8
Interleaver, 57, 58, 59, 60, 63
Irregular LDGM, 96
Iterative decoding, 47, 57, 58, 81

Lattice based Extended Min-Sum, 15, 82, 90, 100
Lattice codes, 2, 15, 67, 69, 81
Lattice constellations, 94, 95, 97
Lattice network coding, 2, 67, 68, 70, 78
Linear programming, 8, 9
Log-likelihood ratio, 3, 34, 58, 83, 105
Lossy wireless networks, 8, 11, 15, 16, 113
Low Density Generator Matrix, 2, 81
Low-energy multicast, 9
Lower bound, 10, 36, 120

Markov chain, 10, 113, 115, 129, 130
Max-flow, 1, 4
Maximum a posterior, 52
Maximum likelihood, 15, 48
Maximum likelihood sequence detection, 15, 48
Medium Access Control, 25
Mesh networks, 4, 7, 13
Minimal mean squared errors, 3
Minimum Mean Square Error, 72, 89, 109
Minimum-energy multicast tree, 9

143

Modified FAST algorithm, 37
Monte Carlo method, 15, 82, 93, 96, 100
Multicast networks, 4
Multicasting group, 13
Multi-hop, 7, 8, 47, 114
Multiple information streams, 6
Multiple interpretations, 14, 15, 25, 27, 32
Multi-source multi-destination, 44
Multi-way, 27, 18, 68, 78, 103
Mutual information, 67, 69, 71, 73, 75

Nearest neighbors, 76, 93
Neighbor coding, 16, 123, 140
Nested coded system, 32
Nested convolutional codes, 68
Nested non-binary, 81, 82, 86, 90, 91
Nested packets, 33
Network layer, 1, 4
Network topology, 6, 8, 14
Normalized Squared Euclidean Distance, 54
NP-hard, 9

One-to-all, 7, 10, 12
Opportunistic scheduling, 14, 25, 27, 28, 32
Optimum distance spectrum, 34
Order statistics, 29
Outage probability, 15, 26, 38, 40, 41

Path, 36, 76, 106, 107, 114
Phase response, 51, 54
Power adaptation, 39, 40
Power efficiency, 6, 9
Power spectral density, 52, 53
Probability
    bit transmission probability, 35
    error probability, 15, 35, 36, 37, 42
Probability Density Function, 28, 52

Punctured Convolutional Code, 49
Puncturing rate, 34

Quadrature phase-shift keying, 108
Quantization error, 69

Rate adaptation, 25, 27, 29, 40, 41
Rate Compatible Convolutional, 32, 48, 53
Rate Compatible Repetition Convolutional (RCRC) codes, 50
RCPC code, 32, 42, 43, 44
Recursive Systematic Convolutional (RSC) encoder, 51
Retransmission, 11, 12, 113, 127
Routing, 1, 7, 8, 9, 13

Scheduling, 14, 25, 27, 28, 32
Shannon capacity, 15, 28
Signal-to-Noise Ratio, 15, 33, 96, 103
Soft information 2, 3, 15, 34, 103
Soft information forwarding, 3, 103, 104
Soft-In Soft-Out, 57
Spectrum efficiency, 2, 67, 81
Store and forward, 1, 7, 47
Superposition, 68, 110

Time varying channel, 29
Trellis coded quantization, 15, 103, 112
Trellis coded quantization/modulation, 15
Two-way relay channels, 15, 103, 105, 107, 109

Undirected networks, 1, 8
Unicast, 7, 8, 47

Variable node, 90, 91
Viterbi algorithm, 106, 107
Voronoi region, 69, 70, 75, 86, 92
Voronoi region of a lattice, 69